AF349020

AI and Robotic Technology in Materials and Chemistry Research

AI and Robotic Technology in Materials and Chemistry Research

Xi Zhu

WILEY-VCH

Author

Prof. Xi Zhu
School of Science and Engineering
The Chinese University of Hong Kong
Shenzhen
China

Cover: © Andriy Onufriyenko/
Getty Images

Library of Congress Card No.: applied for

British Library Cataloguing-in-Publication Data
A catalogue record for this book is available from the British Library.

Bibliographic information published by the Deutsche Nationalbibliothek The Deutsche Nationalbibliothek lists this publication in the Deutsche Nationalbibliografie; detailed bibliographic data are available on the Internet at <http://dnb.d-nb.de>.

Print ISBN: 978-3-527-35428-3
ePDF ISBN: 978-3-527-84882-9
ePub ISBN: 978-3-527-84881-2
oBook ISBN: 978-3-527-84883-6

Typesetting   Straive, Chennai, India
Printing   CPI Group (UK) Ltd, Croydon, CR04YY
C9783527354283_291024

Contents

Preface   *ix*
About the Author   *xi*
Acknowledgments   *xiii*

1   **Survey of Challenges in Chemistry and Materials Science
Research**   *1*
1.1   Introduction   *1*
1.2   Energy Form   *2*
1.2.1   Steam Power   *2*
1.2.2   Electricity Power   *4*
1.2.3   Other Energy Forms   *8*
1.3   Data   *11*
         References   *19*

2   **Robots Technology Development in Modern Scientific
Research**   *21*
2.1   Introduction   *21*
2.2   Early Development of Laboratory Automation (Before 2000)   *22*
2.2.1   Early Automation Technologies   *22*
2.2.2   Laying the Foundation for AI and Robotics   *29*
2.3   Preliminary Integration and Development of Laboratory Automation
         (2000–2019)   *31*
2.3.1   Automation Technologies (2000–2010)   *31*
2.3.2   Various Forms of Exploration Based on Established Foundations   *41*
2.4   Latest Developments and Current Trends (2020–2023)   *42*
2.4.1   Automation Technologies in Five Years   *42*
2.4.2   Mature Industrialization as well as In-depth Exploration   *50*
2.5   Outlook on Future Development   *52*
2.6   Conclusion   *54*
         References   *55*

3   **AI Algorithm for Chemical and Bio-material Design** *57*
3.1   Introduction *57*
3.2   Molecular Representation and Encoding *58*
3.2.1   Linear Notations for Molecules *58*
3.2.2   Graph Representations for Molecules *63*
3.3   The Formulation of Accessible and Searchable Data *66*
3.3.1   Traditional Way for Molecular Structure–Property Relationship Determination: The Kohn–Sham Equation *66*
3.3.2   Dataset Preprocessing *68*
3.3.3   Current Existing Dataset *69*
3.4   AI for Molecular Structure–Property Relationship *71*
3.4.1   The Deep Learning Technology *72*
3.4.2   AI Solving the Kohn–Sham Equation *75*
3.5   AI for Chemical and Bio-material Design *78*
3.5.1   Design Workflows *78*
3.5.2   Example of Designed Chemical and Bio-materials *79*
References *81*

4   **Autonomous Laboratory Empowered by AI and Robotics** *85*
4.1   Evolution of Laboratory *85*
4.2   Core Technologies in Autonomous Laboratories *85*
4.2.1   Autonomous Laboratory Components *85*
4.2.2   Reinforcement Learning *89*
4.3   Example Autonomous Laboratory Solution *90*
4.3.1   Automatic Device only Solution *90*
4.3.2   Solution that Including Design and React *91*
4.3.3   Solution in Reaction Optimization *92*
4.3.4   Solutions Contain Full Phases *94*
4.4   Advanced Autonomous Laboratory Solutions *98*
4.4.1   Advanced Experimental Data Analysis Methods *98*
4.4.2   Design and Analysis in the Large Model Era *104*
4.4.3   Experiment Visualization *106*
4.5   Future Prospects and Trends *108*
References *110*

5   **Large Language Models for the Autonomous Material Research** *113*
5.1   Review of Large Language Models Development and Applications *113*
5.2   Fundamentals of LLM for Material Research: Database and Knowledge Base Construction *121*
5.3   Evaluation: Spider Matrix *124*
5.4   Ideation: AI Supervisor and ScholarNet *130*
5.5   Results and Discussion *134*
5.6   Conclusion *136*
References *136*

6 **Toward a Blockchain-Powered Anti-Counterfeiting Experimental Data System in an Autonomous Laboratory** *139*
6.1 Blockchain Technology *139*
6.2 Laboratory Chemical Management and Safety *143*
6.3 The Problem of Data Integrity and Counterfeiting in Scientific Research *145*
6.4 Blockchain in the Autonomous Laboratory *150*
6.5 Symbolic Representation of Experiments *155*
6.6 Challenges and Limitations *157*
6.6.1 Standard Compilation for Experiment Methods *157*
6.6.2 High Cost for PoW and PoS *158*
6.6.2.1 Data Storage Safety *158*
6.7 Conclusion *159*
References *160*

7 **The Future Integrated Computational and Experimental Research in Metaverse** *163*
7.1 Introduction of Metaverse *163*
7.1.1 Industry 5.0 Protocol *163*
7.1.2 Current Development of Metaverse *164*
7.1.3 Human-in-Loop Paradigm *165*
7.2 Research Paradigm in Metaverse *165*
7.3 Autonomous High-Throughput Experiments *166*
7.3.1 Theory Driven by AI *168*
7.3.2 HIL Implementation *169*
7.3.3 AI Prediction *171*
7.4 H_2O Phase Research in Metaverse *172*
7.5 Aqueous System Research in Metaverse *176*
7.6 Challenges and Future Directions *182*
References *182*

Index *187*

Preface

This book is a systematic summary of my observations of the industry and my own scientific research since I began studying chemical experiment automation at the Chinese University of Hong Kong – Shenzhen in 2017. The research process for many disciplines, including chemistry and materials science, generally consists of two parts: idea generation and idea execution. In traditional chemical and materials research, the three components of theory, computation, and experimentation each have their own unique characteristics, and researchers in each of these areas often have different philosophical views and their own specialized languages for communication. When I was studying for my doctorate in materials computation in Singapore, I would often discuss the theoretical images behind experimental data with my experimental collaborators, and there would often be situations where I thought more experiments should be done, but the other party thought it would be too difficult.

Therefore, since 2010, I have believed that the use of robotics technology in chemical and materials synthesis is a pressing need, especially for those engaged in theoretical and computational chemistry research, as they often need to understand the physical images behind a spectrum from more and denser experimental data, which is beyond the capabilities of human experimental researchers. It was not until July 2016, when I met Professor Yangsheng Xu, the president of the Chinese University of Hong Kong – Shenzhen and a renowned expert in robotics, in the lobby of the Swissotel The Stamford in Singapore, that I expressed my desire to study chemical experiment robots at The Chinese University of Hong Kong (Shenzhen). President Xu quickly expressed his support, and even though I did not have any relevant research experience at the time, I was given the valuable opportunity to conduct independent research in this field in Shenzhen.

Subsequently, I designed the architectures for various intelligent chemical experiment robots, including AIR-Chem (Authentic Intelligent Robotics for Chemistry), MAOS (Materials Acceleration Operation System) and MAOSIC (MAOS in Cloud), BiaeP (Blockchain Integrated Automatic Experiment Platform), and AIM (Authentic Intelligent Machine). The logic behind these is first solve the automation of experimental operations (AIR-Chem), then we need an Internet of Things system to control more robots (MAOSIC), and at the same time, we hope that our experimental data can be anti-counterfeit under the new architecture, so we can introduce blockchain technology (BiaeP), and finally, when we have a sufficient amount of real and effective experimental data, we may be able to challenge some past theories through the Scaling Law approach (AIM).

In fact, our machines cannot execute the ideas of just one or a few people. At the end of 2022, the technology of large language models, represented by ChatGPT, provided an imaginative space for idea generation. Therefore, I have made all of my experimental data, meeting reports, and other textual materials from the past few years, as well as all of my machines, into a large database that can be connected to any large model system, including ChatGPT. In a sense, this is a system that can achieve autonomous research. I am constantly exploring this matter. After the closed loop of idea generation and idea execution is realized, the technology of the digital metaverse may become one of the platforms for the presentation of materials chemistry, which may seem a bit vague and elusive at present, but if we can know the weather conditions around us a few hours in advance, can we also predict chemical reactions? The last few chapters of this book will also provide a detailed introduction to this project.

I often ponder what my contribution to this field is, and I am slowly beginning to understand the answer to this question: I have expanded a simple scalar chemical equation, composed of a few symbols, into a tensor chemical equation that includes the reactants, reaction conditions, reaction vessels, the concentration and performance of the products, and full data visualization of the reaction process. I am working on the architecture of chemical digitization. The chapters in this book are a testament to the progress I have made in this endeavor.

22 September 2024

Xi Zhu
The Chinese University of Hong Kong
Shenzhen

About the Author

Professor Xi Zhu graduated from the University of Science and Technology of China and Nanyang Technological University in Singapore. He is now an associate professor at The Chinese University of Hong Kong, Shenzhen, and has served as the Deputy Director of the Universal Artificial Intelligence Application Research Center at the Shenzhen Institute of Artificial Intelligence and Robotics for Society.

Professor Xi Zhu's research field is the development of AI robotic systems in the experimental science domain that are capable of proposing scientific hypotheses and executing experimental operations. He leads a team that has developed AI-Supervisor, a knowledge analysis and recommendation system in the field of material chemistry, and MAOSIC, a cloud-based material chemistry laboratory utilizing intelligent robots and cloud computing technology. His research focuses on developing a digital system for material science that is led by theoretical frameworks.

Acknowledgments

I am immeasurably thankful for the unwavering support and encouragement of Professor Yi Xie at the University of Science and Technology of China. Since 2005, her guidance in the field of chemical science has been invaluable. My heartfelt gratitude also extends to Professor David Tong and Professor Zhigang Zou at CUHK-Shenzhen. Their continuous backing of my research projects and the wealth of ideas they have shared with me have significantly contributed to my work.

Moreover, I express my deepest appreciation to President Yangsheng Xu at CUHK-Shenzhen. The opportunity to conduct research at this prestigious institution, under his leadership, has been a great privilege. His inspiring ideas in the fields of artificial intelligence and robotics have been a source of motivation for me.

1

Survey of Challenges in Chemistry and Materials Science Research

1.1 Introduction

Chemistry and materials science constitute a profoundly complex and ancient discipline that has faced entirely distinct challenges across different eras. Without delving into the distant past, let us consider the scenario 20 years ago when the author was engaged in undergraduate research within a chemistry laboratory at the University of Science and Technology of China. A formidable challenge at that time was the complete unpredictability of experimental outcomes, which sometimes left the researchers in the dark about the nature of their work. The standard procedure involved mixing prepared solids and liquids in a hydrothermal autoclave, followed by heating in an oven at 180 °C for approximately 24 hours. Subsequently, the mixture was extracted, separated, washed, and prepared for analysis. This involved observations under various electron microscopes to examine the morphology, along with routine completion of other tests, such as X-ray diffraction (XRD) and spectroscopy. Occasionally, tests for lithium-ion battery performance were also conducted. Perhaps one of the most gratifying experiences at that time was observing the artistic beauty of transmission electron microscopy (TEM) images.

Today, we have grown accustomed to the ubiquity of artificial intelligence, big data, and robotics in our daily lives. Looking back at academic papers from the field of chemistry and materials science twenty years ago, especially those concerning nanomaterials, they appear as collections of data interspersed among images, text, tables, and references. During that era, the publication of, or contribution to, an academic paper was often a source of great joy for many. This retrospective underscores not only the dramatic evolution of technology and methodology within the field but also highlights the fundamental nature of scientific inquiry, which remains constant: a quest for understanding and innovation. The transition from manual experimentation and analysis to the integration of advanced computational tools and methodologies has significantly enhanced the capacity for prediction, analysis, and application in materials science. Yet, the essence of discovery, characterized by moments of joy and frustration, the painstaking gathering of data, and the meticulous interpretation of results, continues to define the discipline. This evolution reflects a broader narrative of progress in science and technology, where the accumulation of knowledge and the development of new

AI and Robotic Technology in Materials and Chemistry Research, First Edition. Xi Zhu.
© 2025 WILEY-VCH GmbH. Published 2025 by WILEY-VCH GmbH.

tools mutually reinforce each other, driving the boundaries of what is possible ever forward.

1.2 Energy Form

When we discuss "new energy" today, it is invariably linked to another term, "new materials," and vice versa. The relationship between materials chemistry and energy is one of mutual promotion and complementarity. The generation, storage, transport, and utilization of energy are all reliant on specific functional materials, while more advanced energy systems have enhanced the precision of our observations of the world, significantly propelling the technological progress of materials science. Concurrently, the continuous accumulation of human scientific and technological knowledge further promotes the emergence and application of new technologies. As described by the "materials big data" projects in recent years, combined with the current "generative" artificial intelligence technologies, we seem to have discovered a new domain for more efficient exploration and discovery from existing data toward incremental innovation. Of course, this is predicated on having sufficient computational power, which is itself a part of energy, underscoring the growing importance of technology in the new energy sector. Thus, we observe that today's materials science can be viewed as the process where theory or algorithms drive data through energy to achieve incremental innovation, which represents our primary competitive direction. This improved and expanded version positions the interdependence of new energy and new materials within a broader scientific and technological context, emphasizing the role of computational power and artificial intelligence. It sets the stage for a detailed historical analysis, hinting at the evolution of these fields and their impact on contemporary scientific research and technological development. I will proceed to analyze this process from a historical perspective.

1.2.1 Steam Power

In analyzingthe trajectory of materials chemistry within the broader context of societal development and energy paradigms, it becomes evident that the evolution of this field is deeply intertwined with the predominant energy sources of its respective eras. The progression from a society reliant primarily on human and animal labor to one powered by steam, and eventually to our current age of electricity and emerging renewable energies, has had profound implications for the advancement of materials chemistry.

During the pre-industrial era, characterized by manual labor, the field of materials chemistry was in its nascent stages. The absence of sophisticated instrumentation and analytical techniques meant that researchers' understanding of chemical phenomena was limited to observable reactions and processes that could be achieved without the aid of advanced technology. This period's knowledge base was foundational yet primitive by today's standards, focusing on the basic properties of materials and their simple transformations.

Figure 1.1 The revolutionary invention of the steam engine marked a monumental leap from manual labor to mechanized production, symbolizing a pivotal moment in human history. The Steam Age stands as a crucial milestone in human history, catalyzing industrialization and modernization, reshaping production methods, social structures, and lifestyles, exerting profound and enduring influence on the world.

The Industrial Revolution marked a pivotal shift, with the invention and widespread adoption of the steam engine catalyzing an unprecedented expansion in industrial capabilities and scientific inquiry. The steam engine, a marvel of engineering and materials science, necessitated the development of materials that could withstand high pressures and temperatures. This requirement spurred significant advancements in metallurgy, exemplified by the Bessemer process, which revolutionized steel production by making it more efficient and cost-effective. The ability to produce stronger, more durable materials was not just a technological achievement but also a cornerstone in the edifice of modern industrial society, enabling the construction of railroads, bridges, and machinery that powered the nineteenth century's economic expansion (Figure 1.1).

Furthermore, the steam era's influence extended into the realm of chemical production and analysis. The coal industry, a key driver of the steam engine, became a vital source of raw materials for the burgeoning chemical industry. Coal tar, a byproduct of coal gasification, was the precursor for an array of chemical dyes, initiating a new era in the textile industry and laying the groundwork for synthetic organic chemistry. The development of analytical chemistry was equally crucial, with innovations such as spectroscopy and chemical thermodynamics emerging in response to the industrial and scientific challenges of the time.

The establishment of dedicated research institutions and the systematic approach to materials chemistry research were also hallmarks of this era. The professionalization of chemistry as a distinct scientific discipline, coupled with the enhanced collaboration between scientists and engineers, led to a more methodical and empirical approach to research. This collaborative ethos was instrumental in bridging the gap between theoretical chemistry and its practical applications, fostering a culture of innovation that would pave the way for the next century's scientific breakthroughs. The steam power era's legacy is its role in promoting the global spread of chemical knowledge. The advent of steam-powered printing presses made scientific literature more accessible, while improved transportation facilitated the exchange of ideas and materials between researchers across the globe. This era laid the foundational principles of materials chemistry as we understand it today, setting the stage for the subsequent development of polymers, composites, and nanomaterials that are essential to modern technology.

As steam technology advanced, scientists gained a deeper understanding of thermodynamics, marking a period of significant progress in the field. This era was also characterized by burgeoning theoretical research in reaction kinetics, reflecting an increasing sophistication in the comprehension of the forces and principles governing chemical reactions. Concurrently, the chemical engineering industry experienced sustained growth, driven by these scientific advancements and the demand for industrial applications of chemical processes. In parallel, the field of reaction kinetics emerged, focusing on the rates at which chemical reactions occur and the factors influencing these rates. This area of study is vital for understanding how reactions can be optimized for industrial processes, including those used in the chemical engineering industry. Theories related to reaction kinetics, such as the Arrhenius equation (Arrhenius 1889), which describes how reaction rates increase with temperature, became instrumental in the design and improvement of chemical reactors and processes.

The expansion of the chemical engineering industry during this time can be attributed to the integration of these scientific insights into practical applications. Chemical engineers leveraged the principles of thermodynamics and reaction kinetics to develop processes that are more efficient, cost-effective, and capable of producing materials and chemicals at a larger scale. This not only facilitated the growth of the chemical industry itself but also had a wide-reaching impact on sectors such as pharmaceuticals, energy, and materials science, contributing to the advancement of society as a whole. Thus, the advancement of steam technology and the deepening understanding of thermodynamics and reaction kinetics played pivotal roles in the scientific and industrial growth of the 19th and early 20th centuries, marking a period of remarkable innovation and expansion in the chemical engineering field.

1.2.2 Electricity Power

The transition from steam to electric power marked a revolutionary period in human history, heralding the second industrial revolution. This era, spanning the late 19th and early 20th centuries, was not just about the adoption of electricity as a primary energy source but also about the profound impact it had on materials chemistry. The electrification of society demanded new materials with specific properties, fostering a wave of innovation in chemistry and material science.

An understanding confined solely to gases is significantly inadequate for the field of materials chemistry, which also heavily involves the study of condensed phases such as solids and liquids. These phases arguably represent a more prevalent subject of research. In the domain of solid-state physics, electrons emerge as one of the pivotal research subjects. The study of electrons is indissolubly linked to the understanding of electricity and the socio-economic context of the era. By the late nineteenth century, the exploration and application of electricity had made considerable strides, with the advent of technologies like the light bulb, telephone, and electrical power distribution witnessing a qualitative leap and achieving widespread practical application. Many scientists of that period began to leverage electricity as a tool for scientific investigation. For instance, in 1895, Röntgen (1896) discovered X-rays while experimenting with a Crookes tube – an early experimental electrical discharge tube – during his studies on cathode rays. This discovery underscored the pivotal role of electrical technology in facilitating scientific breakthroughs.

The unveiling of X-rays opened the door to observing the microscopic structure of materials, further broadening the horizons of material chemistry through the advent of quantum mechanics. Quantum mechanics has introduced concepts such as momentum space into contemporary solid-state physics, significantly expanding our cognitive and exploratory scope in materials chemistry. This progression underscores the interdisciplinary nature of materials science, highlighting how advances in one area can propel understanding and innovation across multiple scientific domains. It exemplifies the profound impact of electrical studies and technological advancements on the evolution of materials chemistry, enabling the detailed examination of material properties at the atomic and molecular levels. Consequently, these insights have paved the way for the development of new materials and technologies, underscoring the integral role of a comprehensive understanding of all states of matter – solid, liquid, and gas – in the advancement of materials science and engineering (Figure 1.2).

With the maturation of solid-state physics (Ashcroft and Mermin 1976) and the conceptual framework of momentum space, the intricate relationship between

Figure 1.2 The advent of electricity ushered in an era of unprecedented innovation and connectivity, fundamentally transforming human civilization and laying the groundwork for the modern technological landscape. The Electrical Era represents a paradigm shift in human history, sparking revolutions in communication, transportation, and industry, fostering global interconnectedness, and fostering the birth of countless inventions that continue to shape our daily lives.

structure and properties has been catapulted to the forefront of tangible research inquiries, particularly in the realms of electrical conductivity and band gap analysis. Solid-state physics, as a discipline, seeks to elucidate the physical properties of solids from a microscopic perspective, leveraging quantum mechanics as a fundamental theoretical underpinning. This approach has profoundly enriched our understanding of how the atomic and electronic structures of materials influence their macroscopic properties.

The concept of band theory, a cornerstone of solid-state physics, offers a comprehensive explanation for the electrical conductivity of materials. Band theory posits that the energy levels of electrons in a solid form bands of energy rather than discrete levels. The presence of band gaps, or energy ranges in which no electron states can exist, plays a crucial role in determining a material's conductivity. For instance, insulators are characterized by a wide band gap, preventing electrons from easily moving across the energy barrier, whereas conductors exhibit little to no band gap, facilitating free electron movement and, consequently, electrical conductivity. Semiconductors, pivotal in modern electronics, possess a narrow band gap that allows their conductive properties to be finely tuned through doping or environmental changes.

Moreover, the advent of quantum mechanics has enabled the prediction and manipulation of material properties with unprecedented precision. Quantum mechanics (Griffiths and Schroeter 2018) introduces the principle of wave-particle duality, which asserts that particles such as electrons exhibit both particle-like and wave-like properties. This principle is instrumental in understanding the quantum mechanical behavior of electrons in solids, which directly influences material properties such as electrical conductivity, magnetism, and optical absorption.

The exploration of momentum space, a quantum mechanical construct where positions and momenta are conjugate variables, has further deepened our comprehension of solid-state phenomena. The analysis of electronic states within momentum space facilitates a more nuanced understanding of band structures and electron dynamics, contributing to advancements in material design and application. These scientific advancements underscore the crucial role of solid-state physics in the modern technological landscape. The ability to engineer materials with tailored electrical, optical, and magnetic properties has been a driving force behind the development of advanced technologies, ranging from semiconductor devices and solar cells to quantum computing. The ongoing research into the structure–property relationship not only enriches our theoretical knowledge but also paves the way for the innovation of new materials and technologies that address the complex challenges of the twenty-first century.

During the steam era, the focus was largely on macroscopic physical properties and the statistical behaviors of gases and vapors. This period was marked by the development of thermodynamics, which provided a framework for understanding energy conversion and efficiency in steam engines and other heat-driven systems. The limitations of steam power, however, were evident in its reliance on bulky machinery, the inefficiency of heat engines, and the localized nature of power generation and distribution. The advent of the electrical age represented a quantum leap forward, not just in terms of energy production and utilization, but also in the granularity and precision of scientific research. Electricity offered an unprecedented level of control and versatility, enabling the study of individual atoms and molecules and the electronic properties of materials. This shift was pivotal for the field of materials chemistry, where the focus expanded from the collective behavior of particles in gases to the intricate details of solid materials, including their atomic and electronic structures.

Moreover, the electrical age has brought about significant advancements in energy sustainability and efficiency. Unlike the steam age, which was heavily dependent on coal and other fossil fuels, electrical power can be generated from a variety of sources, including renewable energy such as solar, wind, and hydroelectric power. This diversification of energy sources is crucial for addressing the environmental challenges of the twenty-first century, highlighting the role of electrical power not only in advancing scientific knowledge and technological capabilities but also in promoting sustainable development.

It's essential to emphasize that in the electrical age, the demands of the industrial sector have significantly propelled advancements in materials chemistry. This symbiotic relationship between industrial needs and scientific innovation has led to remarkable progress in materials development, directly influencing the efficiency, sustainability, and capabilities of modern technologies. The industrial demands for better performance, reduced costs, and enhanced sustainability have been key drivers in the evolution of materials chemistry. Industries reliant on electrical technologies, such as electronics, automotive, aerospace, and energy, have continuously pushed the boundaries of what's possible with current materials, urging scientists to innovate and develop new materials with superior properties.

In the electronics industry, for instance, the relentless pursuit of miniaturization and higher performance has driven the development of advanced semiconductor materials beyond silicon, including gallium arsenide (GaAs) and graphene. These materials offer superior electrical conductivity and electron mobility, enabling faster and more energy-efficient electronic devices. The quest for more efficient photovoltaic cells has similarly spurred research into novel materials that can convert sunlight into electricity more efficiently, such as perovskite solar cells, which offer a promising alternative to traditional silicon-based cells with the potential for higher efficiency and lower manufacturing costs.

The automotive and aerospace industries have also been instrumental in advancing materials chemistry, particularly in the development of lightweight and high-strength materials. The shift toward electric vehicles (EVs) and the need for longer battery life have intensified research into advanced battery technologies, including lithium-ion batteries with improved energy density and charging speeds. Materials such as carbon fiber composites and aluminum alloys have become crucial for reducing vehicle weight, enhancing fuel efficiency, and improving performance.

Moreover, the energy sector's shift toward renewable sources has catalyzed the development of materials that can efficiently capture, store, and convert energy. Innovations in materials chemistry have led to more efficient wind turbines, safer nuclear reactors, and more durable hydroelectric power facilities. The advancement of energy storage technologies, including supercapacitors and next-generation batteries, is critical for addressing the intermittency of renewable energy sources and ensuring a stable and sustainable energy supply.

This industrial push for innovation has not only led to the development of new materials but has also necessitated advancements in materials characterization and manufacturing techniques. Techniques such as atomic layer deposition, 3D printing of functional materials, and advanced microscopy and spectroscopy methods have evolved to meet the intricate demands of materials synthesis and analysis, enabling the precise engineering of materials at the atomic and molecular levels.

1.2.3 Other Energy Forms

Following the unprecedented technological progress made through the first two industrial revolutions, the field of materials chemistry has seen significant advancements. Subsequently, there emerged a variety of new energy forms, including nuclear power, lithium-ion batteries, hydrogen energy, and even renewable sources such as hydroelectric and wind power. Cleanliness and environmental sustainability have become the new benchmarks for future energy sources. However, from the perspective of researchers in the field of materials chemistry, the ultimate forms of new energy, regardless of the specific application scenarios, remain thermal and electrical energy. Although fundamentally, new energy sources have not revolutionized the basic modalities of materials science research from a theoretical or experimental standpoint, the development of new energy and its related industries has propelled significant growth in the discipline of materials chemistry, enabling a synergistic empowerment with other societal sectors.

The advent of these new energy sources has introduced complex challenges and opportunities for materials chemistry. The development and optimization of materials for nuclear reactors, for example, require an intricate understanding of radiation resistance, thermal conductivity, and mechanical strength. Similarly, the proliferation of lithium-ion batteries has spurred extensive research into electrode materials, electrolytes, and separators to improve energy density, charge rates, and safety profiles (Figure 1.3).

The development of both renewable and nuclear energy has significantly advanced the field of materials chemistry by creating a demand for new materials with unique properties. These materials are tailored to efficiently harness solar, wind, geothermal, and other renewable sources, as well as to withstand the demanding conditions of nuclear reactors. This dual drive toward sustainable and powerful energy solutions has led to a surge in research and innovation within materials chemistry, focusing on improving energy conversion, storage, and transmission.

Renewable energy development has promoted materials chemistry through the quest for more efficient and cost-effective solar panels, leading to research into novel photovoltaic materials like thin-film technologies, organic, and perovskite solar cells. In wind energy, advancements have been crucial in developing stronger, more durable materials for turbine blades to increase efficiency and lifespan.

(a) (b)

Figure 1.3 (a) Renewable energy: embracing a diverse array of sources like wind, hydro, and solar, renewable energy stands at the forefront of the global shift toward sustainable practices. It offers a promising avenue for reducing greenhouse gas emissions and minimizing our ecological footprint. As a key player in combating climate change, renewable energy not only contributes to a cleaner environment but also supports economic stability by creating jobs and reducing dependency on fossil fuels, encapsulating the dynamic progress and potential of sustainable development. (b) Nuclear power: with its immense potential and controversial implications, nuclear energy represents a double-edged sword, offering vast amounts of carbon-free power while posing significant challenges in terms of safety, waste management, and proliferation concerns, highlighting the complexities of navigating our energy future.

The integration of renewable sources into the power grid has also necessitated advancements in energy storage technologies, with materials chemistry playing a pivotal role in developing advanced electrolytes and electrode materials.

Photovoltaic Materials for Solar Energy. The quest for more efficient and cost-effective solar panels has spurred research into novel photovoltaic materials. Beyond traditional silicon-based solar cells, materials chemists have been exploring thin-film technologies, organic and perovskite solar cells, aiming to increase efficiency, reduce costs, and offer flexible, lightweight options for solar energy generation. This research directly responds to the renewable energy sector's demands for more versatile and efficient solar technologies.

Materials for Wind Energy. Advancements in materials chemistry have been crucial for the wind energy sector, particularly in developing stronger, more durable materials for wind turbine blades. Research into composites and polymers aims to create blades that are not only lighter and stronger but also capable of withstanding harsh environmental conditions, thereby increasing efficiency and lifespan of wind turbines.

Electrolytes and Electrodes for Energy Storage. The integration of renewable energy sources into the power grid has necessitated advancements in energy storage technologies, such as batteries and supercapacitors. Materials chemistry plays a pivotal role in developing advanced electrolytes and electrode materials that offer higher energy densities, faster charging times, and longer life cycles. Innovations in lithium-ion batteries, solid-state batteries, and beyond are directly driven by the needs of the renewable energy sector.

Materials for Hydrogen Production and Fuel Cells. The shift toward hydrogen as a clean energy carrier has promoted research in materials chemistry for efficient hydrogen production, storage, and utilization in fuel cells. Developing catalysts for water electrolysis, materials for hydrogen storage, and proton-exchange membranes for fuel cells are critical areas where materials chemistry contributes directly to advancing hydrogen as a renewable energy source.

Thermoelectric and Piezoelectric Materials. Renewable energy development has also accelerated research into thermoelectric and piezoelectric materials, which convert waste heat and mechanical energy into electricity, respectively. These materials offer potential for energy harvesting in a variety of settings, contributing to the efficiency and sustainability of energy systems.

Similarly, the development of nuclear power has necessitated the creation and improvement of materials capable of handling the extreme conditions of nuclear reactors, such as high temperatures and radiation. This need for specialized materials has spurred substantial research and innovation in materials chemistry, focusing on enhancing the safety, efficiency, and longevity of nuclear energy systems.

Radiation-Resistant Materials. The operation of nuclear reactors involves exposure to intense radiation, which can degrade many materials over time. This challenge led to the development of radiation-resistant materials, including specific alloys and ceramics that can maintain structural integrity and functionality in high-radiation environments. Research in understanding how materials interact with radiation has been a direct outcome of the nuclear power industry's requirements.

High-Temperature Materials. Nuclear reactors operate at high temperatures, necessitating materials that can withstand these conditions while maintaining strength and corrosion resistance. This requirement has driven advancements in high-temperature materials science, including the development of superalloys and advanced ceramics that can perform under the extreme thermal conditions found in reactors.

Fuel Cladding Materials. The development of materials for fuel cladding, which encases the nuclear fuel to prevent the release of radioactive particles, is another area where nuclear power has propelled materials chemistry. Innovations in zirconium alloys, for example, have been critical in creating effective fuel cladding that minimizes corrosion and allows for efficient heat transfer.

Coolant and Moderator Materials. The search for efficient coolant and moderator materials, which play crucial roles in the operation of nuclear reactors by managing reactor temperatures and neutron flux, respectively, has led to significant research in materials chemistry. This includes the development of liquid metals, gases, and graphite with specific properties tailored to nuclear applications.

Waste Management and Containment. The handling and containment of nuclear waste have necessitated advancements in materials capable of safely isolating radioactive materials from the environment for extended periods. This challenge has led to innovations in containment materials, including glass, ceramics, and cements, designed for long-term stability and resistance to radiation and chemical degradation.

Both the renewable and nuclear energy sectors have served as catalysts for significant advancements in materials chemistry, driving the development of new materials that enhance the efficiency, sustainability, and safety of energy technologies. This symbiotic relationship highlights the critical role of materials chemistry in powering the future with a blend of sustainable and reliable energy sources, showcasing the field's extensive contributions to addressing global energy challenges.

These advancements are part of a broader trend where materials chemistry is pivotal in tackling the energy challenges of the twenty-first century. The discipline not only provides new materials but also offers insights into the mechanisms at play within these materials, enabling the design of systems that are more efficient, sustainable, and environmentally friendly. Additionally, the intersection of materials chemistry with other industries has led to a fruitful exchange of ideas and technologies. For instance, developments in photovoltaic materials influence the automotive industry by enabling the creation of lighter, more energy-efficient vehicles. Similarly, advancements in materials for energy storage have significant implications for both the grid and transportation sectors, potentially revolutionizing energy consumption patterns.

In summary, while new energy forms have not fundamentally changed the theoretical or experimental foundations of materials science research, they have greatly expanded the field's scope and impact. The evolution of materials chemistry in response to new energy challenges exemplifies the dynamic nature of scientific inquiry, where advancements in one domain catalyze progress across multiple areas. This interplay between materials chemistry and new energy technologies underscores the field's crucial role in fostering a sustainable and technologically advanced future.

1.3 Data

The advancement of science and technology has significantly increased the precision of experimental testing, fundamentally improving our understanding of the relationships between structure and properties in materials chemistry. Enhanced experimental techniques, such as high-resolution microscopy and sophisticated spectroscopy, allow for detailed observations of materials at the atomic level. These insights are crucial for identifying how structural nuances influence material properties.

Furthermore, the integration of computational models with experimental data accelerates the exploration of new materials, enabling predictions about how changes in structure affect properties before physical experiments are conducted. Techniques like chemical vapor deposition (CVD) and atomic layer deposition (ALD) exemplify the precise control over experimental conditions necessary for synthesizing advanced materials with specific desired properties.

This improvement in experimental precision not only propels basic research but also drives technological innovations, such as more efficient solar panels and batteries with higher energy densities. As experimental methods continue to evolve, they promise to unveil new phenomena, enhancing our understanding of materials and opening up further possibilities for innovation. In essence, the progress in experimental accuracy enriches our material knowledge, catalyzing advancements across both scientific and technological landscapes.

From the era of steam engines to the electrical age, humanity has undergone a transformation in the way we harness energy, thereby accelerating the processes of material production and synthesis. This has naturally accumulated a vast amount of knowledge, transmitted through education across generations. However, due to various reasons, it is inevitable that knowledge will be lost or even inaccurately transmitted during the educational process. For example, the question of what a "dragon" looks like in Chinese mythology is difficult to answer definitively. Therefore, with the widespread adoption of computer hardware technology, transforming knowledge into data and storing it in hard drives can preserve it for many years. The key concepts of "data" and "many years" lead us to the fundamental characteristic of true technology: the ability to traverse time or space. We can express this in more complex terms using the language of physics: the implications of Noether's theorem (Noether 1918).

To illustrate this with a familiar example, consider the camera technology we are now very familiar with, which allows us to see photographs and images from decades ago. Similarly, telephone communication technology enables us to engage in real-time communication with friends thousands of kilometers away or even in space stations, disregarding spatial distance. Nowadays, camera and telephone technologies can be seamlessly integrated into a single smartphone, which is why companies excelling in this field, such as Apple, have become among the world's most profitable companies – a fact that should come as no surprise.

Let's consider the development of materials chemistry and the potential challenges it may face using the logic of time and space traversal. Firstly, chemicals themselves can be a part of smartphones, which is enough to demonstrate the importance of materials chemistry. This conclusion is correct, but the logic may not be. Currently, we generally believe that the emergence of products like smartphones has gone through a progression from non-smart to smart phones. The competitiveness of these products is not solely driven by their material composition. In simpler terms, the logic behind smartphone development is not primarily dominated by materials science.

Take another example, the pharmaceutical industry. The significance of high-quality drugs does not need overstating, and compared to the smartphone industry, the connection to chemistry is more intricate here. In a specific drug development process, numerous chemical molecules are tested based on demand, which is a complex and enduring process. Each chemical molecule in any testing stage may be proven unable to progress to the next stage. There are many blind spots in clinical stages, and selection can only be made based on experimental results. Therefore, drug development is a high-investment, high-risk industry. Thus, whether in the smartphone or pharmaceutical industry, chemical materials are objects of selection rather than having the power of active choice. In simple terms, many industries require the properties of chemical materials, but these properties cannot be mapped directly to structure, necessitating a lot of experimentation. This extensive experimentation itself consumes a lot of time and resources, and even after numerous attempts, the expected results may not be achieved. Meeting the expected results at one stage does not necessarily mean meeting the expected results at the next stage, leading to wasted efforts. Understanding this, you might understand why materials chemistry is not a popular major in universities, especially when a large number of repetitive experiments currently rely on human labor, with low success rates after time investments. Of course, there are also factors such as high personal safety risks and high degree of repetition.

From the described phenomena above, two main problems arise:

1. The ambiguous relationship between structure and performance.
2. Lengthy testing cycles for experimental data.

We see that these are still problems related to time and space. The corresponding solutions are, then:

1. Clarifying the complex relationship between structure and performance.
2. Shortening the testing cycle.

These solutions might seem like they're saying everything and nothing at the same time. Indeed, many problems have similar answers. The answer to the first problem is knowledge, while the answer to the second problem is execution. Interpreted through ancient Chinese wisdom, this is "unity of knowledge and action."

Let's first explore the realm of knowledge. We've elaborated on the journey from the era of steam engines to the current electrical and new energy era over the past 200 years. Throughout this process, the discipline of materials chemistry has been present and has accumulated a vast amount of knowledge and technology. Objectively, it has also promoted the development of other disciplines. However, the primary contributors and holders of this knowledge and technology have been the traditional developed countries such as Europe, North America, and Japan. Most of the powerful materials chemistry giants undoubtedly hail from these countries.

Now, let's briefly examine the situation of materials chemistry in China. Since joining the World Trade Organization in 1999, China's chemical industry has undoubtedly become one of the most important components of the global chemical industry. Chinese universities and research institutes have surpassed the United States in publishing academic papers in the field of chemical materials, becoming the world leader. It took less than 30 years for China to catch up from a relatively backward state, which could not have been achieved without mutual learning and communication with countries worldwide. The most significant aspect we observe is the tremendous change brought about by the transmission of knowledge. We can roughly estimate that China has learned and mastered approximately 200 years' worth of knowledge in just 30 years, leading to enormous economic and academic growth.

Let's carefully contemplate the two numbers and the logic behind them. The process of these 30 years is driven by continuous human effort, whether it's physical laborers or academic workers contributing. The most typical data point is the significant increase in urbanization rate and the expansion of universities, both in terms of the number of students and research personnel. In line with our previous discussion, this is a chemically driven process. However, the essence of the knowledge gained over these 200 years is essentially stock knowledge. In terms of incremental knowledge, cutting-edge biotechnological materials technology still needs to be imported from Western countries. Here, we would not go through the international geopolitical and economic issues. If we merely consider China's materials chemistry industry as a model, we can summarize its characteristics: acquiring a vast amount of knowledge in a short period and naturally predicting its future demand, which is to acquire even more knowledge in a shorter time. It's essential to note that here we are referring to "knowledge," not just "technology."

Let's take a look around. In today's electrified age, a large number of factories, including chemical plants, produce various chemical industrial products in large quantities through automation technology. One reason for this mass production is that mechanization and automation have replaced manual operations with standardized procedures, thereby significantly increasing the number of finished products per unit time and reducing costs. Robots and automation technology are widely used in the industrial sector, enhancing production efficiency and product quality. However, their application in chemical, materials, and biological laboratories is relatively limited. Why is that?

The reason lies in the stark differences between technological implementation in industry and knowledge realization in academia, in terms of quantity and process requirements. One involves the systematic digestion of existing knowledge, while the other entails free exploration for incremental transformation. Thus, even though chemical plants around us started automating and digitizing much earlier than 50 years ago, and despite having more and better-quality experimental testing equipment, the layout of chemical materials laboratories has not changed significantly. As shown in the images below, on the left is a picture of a chemistry laboratory at Beijing Institute of Technology in 1960, and on the right is a picture of a chemistry laboratory at a certain university in 2024. Both images show rows of experimental workbenches, cabinets underneath, and many chemical reagents and drugs on the workbenches. Despite spanning 64 years, there is no significant difference between the two images. We still rely on our graduate students and postdocs to operate chemical containers and obtain experimental data, and the entire process of chemical experimentation has not achieved automated and intelligent assembly-line-style operation (Figure 1.4).

Indeed, we can argue that the complexity and variability of laboratory work, along with the need for flexibility and specialized knowledge, make it challenging to achieve automation. However, from a solution perspective, there is not much difference compared to solving other physical chemistry problems. We start by addressing one point, then move on to handling the relationships between this point and others, and gradually address the relationships between more points. Although the reagent bottles used in chemical experiments are standardized – for example, in my laboratory, we only have three types: 10, 100, and 500 ml bottles – we cannot demand standardization when it comes to the actions of our students. For

(a)

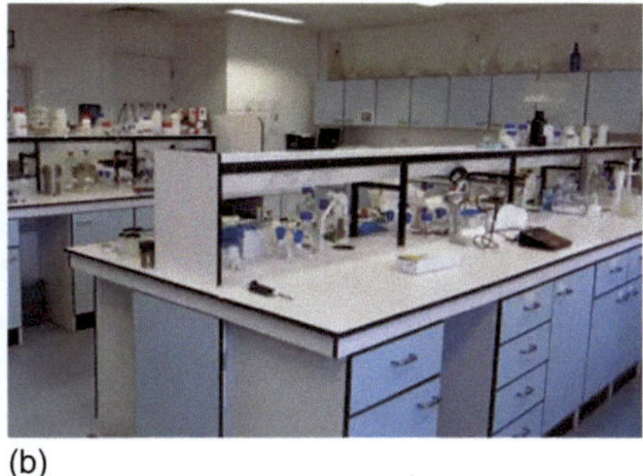
(b)

Figure 1.4 The chemistry lab in (a) 1960s (SinaEducation 2010) and (b) 2024.

instance, we cannot insist that they must assemble a 100 ml reagent bottle within one minute, and that the turbulence bubbles in the solution must be uniform. While the latter may be difficult to achieve, the former could be easily realized through automation technology. However, it may not seem necessary to limit the assembly time to one minute until we realize that the time for reagent preparation may have subtle connections with the quality of the final product.

There is another crucial logical issue here. In the history of chemical materials development, many significant discoveries have occurred due to accidentally using the wrong reactants or incorrect amounts of reactants. For example, the discovery of conducting polymers, which began with Hideki Shirakawa's accidental discovery in 1971 (Shirakawa and Ikeda 1971). At the time, he was supervising a group of South Korean exchange students conducting experiments on the synthesis of polyacetylene. Perhaps due to a language barrier, the students mistakenly increased the amount of the designated catalyst, the Ziegler–Natta catalyst, by thousands of times. As a result, they produced a silver-colored polyacetylene film, overcoming the challenge of synthesizing highly crystalline polyacetylene. The lightweight structure obtained provided an excellent foundation for future doping to enhance electrical conductivity. If Shirakawa had used a machine to precisely conduct the experiment, such an important discovery might not have been made. Although this is a relatively specific example, it is certainly not an isolated case. Historically, laboratory practitioners seem to not prefer automated equipment and consider experimental operations as an integral part of the experiment itself. There are two specific issues here.

Firstly, many researchers do not handle multiple experimental tasks and requirements simultaneously. In the same period, many experimenters often only need to focus on a specific project. This may be determined by the project's budget, timeline, and the current research paradigm. Conversely, pharmaceutical companies have different practices. Over 30 years ago, many pipetting workstations were already in operation. Although these workstations generate vast amounts of research data every day, when I deployed such a workstation at a Chinese university in 2023, I found that the doctoral students there did not frequently use the machine. One reason is that they could not handle so many samples at once.

We have learned about the significant differences between university laboratories and corporate laboratories, as well as the reasons why it is more challenging for university laboratories to adopt intelligent and automated equipment: the conflict between academic credit and task division. In a corporate setting, each position has very detailed task division, so there are clear application scenarios for accelerating processes through automation equipment. However, university laboratories are different. Academic credit for research achievements is a crucial indicator, so it is not feasible to distribute tasks or outsource them through assembly-line-style operations. Researchers theoretically need to be familiar with all aspects of the research process, especially many sequential steps. Researchers need to understand detailed information about all aspects to design and correct the next experimental operation. Current experimental automation equipment focuses on the needs of pharmaceutical companies and other discrete experiments within enterprises but does not provide intelligent equipment with a full-process information architecture. Therefore, for specific chemical materials researchers, the use of local automation acceleration equipment, including pipetting workstations, does not significantly promote research efficiency. Therefore, the main reason why the current paradigm of chemical material science research, mainly led by universities and research institutes, has not achieved a paradigm shift is that there is no automation stage, and it must directly enter intelligence.

I mentioned earlier that in the future, whether in China or the world, the chemical materials industry or any other industry will need to master more knowledge in a shorter time. The above paragraph explains why "intelligent" equipment, rather than "automatic" equipment, is the solution for "shorter time." Next, we need to explain what the solution for "more knowledge" is. Suppose a laboratory researcher synthesizes a chemical molecule that contributes to the treatment of a certain disease. Is this molecule knowledge? Of course, it is, but it is far from enough. The full process information including reaction conditions, synthesis pathways, and various experimental details involving this molecule is more critical knowledge, and the amount of knowledge is closely related to the amount of information. Here, we need to mention the difference between patents and papers. In order to protect the intellectual property related to the knowledge, a patent related to molecule synthesis often discloses many process details, which are protected commercially, but still promote knowledge dissemination. Authors and readers of academic papers often focus more on the conclusions themselves and may not deliberately detail the implementation process because they care more about personal academic credit, and academic papers themselves are not a form of protectable intellectual property. Of course, there are many other problems in the academic paper field, such as unreliable conclusions, poor reproducibility, repetitive research, academic integrity, and so on, which can seriously hinder or even mislead knowledge dissemination. There are countless similar events, and dealing with these problems is very difficult. For example, an American scientist misled the work of 100 research groups in China for 10 years with a false paper, and there is currently no organization that can handle such incidents. In comparison, patents are much simpler, and problems can be solved through domestic or international legal means (Figure 1.5)

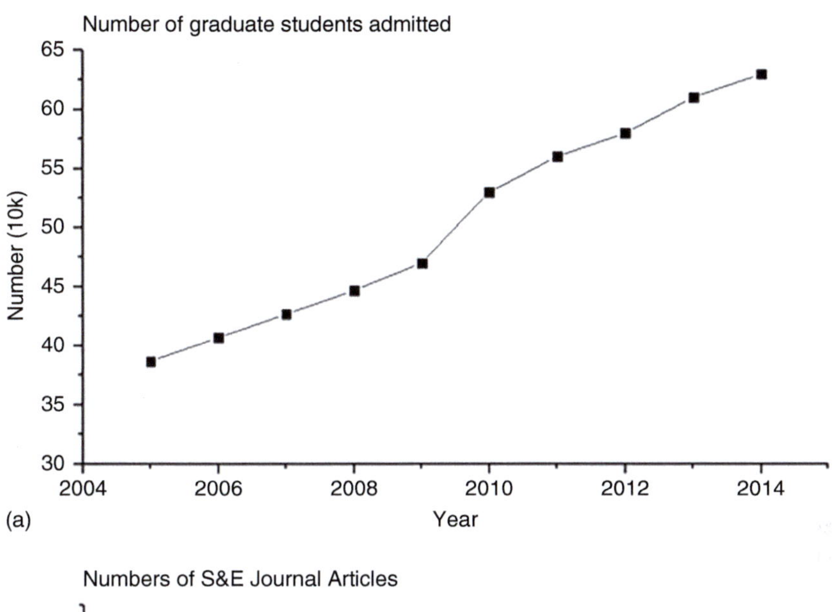

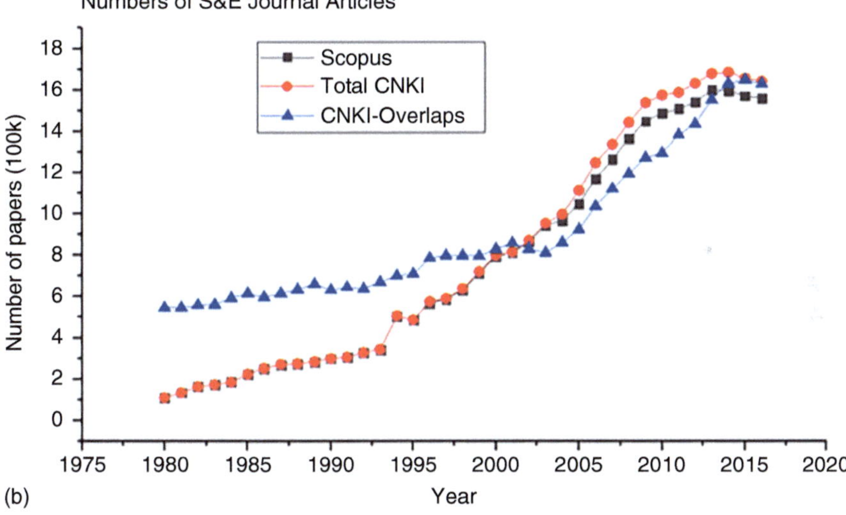

Figure 1.5 (a) The number of graduate students in China has grown since 2004 by about 80% by 2014 (ChinaEducationOnline 2015). (b) The numbers of S&E Journal Articles in Scopus and in CNKI, 1980–2016 (VoxChina 2018).

Let's discuss the issue of the rate of knowledge generation using academic papers as an example. A report (Miyazaki 2023) by *Nature* journal in 2023 stated that China had surpassed the United States in the number of high-quality journals indexed in the Nature Index, with the number of journals in the field of chemical materials surpassing even earlier. There are certainly many reasons why China has achieved this surpassing. Let's take a look at the two charts above. We can observe that the number of graduate students enrolled in China has increased from approximately 360,000 in 2005 to about 640,000 in 2014, an increase of about

80% (ChinaEducationOnline 2015). Meanwhile, the number of scientific and technological papers has increased from about 800,000 in 2005 to about 1.6 million in 2014, a growth of about 100% (VoxChina 2018). We notice that these two numbers are very close, indicating that the increase in the number of papers in China cannot be separated from the expansion of the number of graduate students, including in the field of chemical materials science. If we assume that, on average, an academic paper in the field of chemical materials represents the synthesis or discovery of a new material, then is the rate of discovering new materials meaningful?

Actually, it is not meaningful because it is still too slow! We can do a simple and rough calculation to estimate how many chemical experiments we can conduct. Figure 1.6 summarizes and classifies the total ~2 million papers published in the field of materials science in 2022 through our KnowledgeWorks (Yanheng et al. 2024), divided into 15 categories based on application scenarios, including Energy Storage and Conversion, Catalysis, Environmental Applications, and other familiar categories. Each application scenario can be achieved through a maximum of eight experimental methods. Let's assume there are a total of 1 million reagents, 10 reaction steps, 100 solvents, and up to 4 solvent mixtures, as well as approximately 100 reaction conditions for each of the 10 reaction steps. Calculating this, we find:

$$15 \text{ Applications} \times 8 \text{ Methods} \times (10^6 \text{ Reagents})^4 \times (10^3 \text{ Conditions})^{10} > 10^{100}$$

This means that exploring all known chemical reactions would require a dataset of one Googol (10^{100}), while the amount of data currently explored by humanity is probably less than 10^{10}. Perhaps under the current research paradigm, we can still make many significant discoveries, but the overall conquest still seems far out of reach (Figure 1.6).

This is the biggest challenge in the field of chemical materials science. Currently, it is still driven by human chemistry, much like agricultural societies, where the production of agricultural products is almost entirely linearly dependent on the number of laborers, and a country's overall strength is closely related to its population. Therefore, we often see in some countries that pursue scientific research outcomes, many research teams with hundreds of members, most of whom are graduate students. If we liken the discovery or synthesis of new chemical substances to a treasure hunt

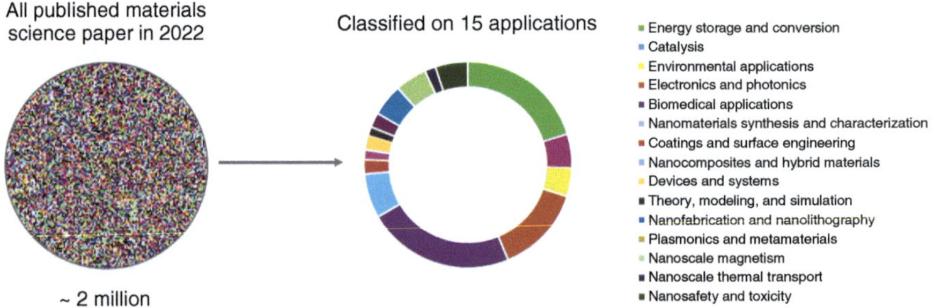

Figure 1.6 There have been ~2 million papers published in materials science since 2022. The papers can be classified according to applications into 15 categories, as shown in the figure. They can also be classified to 8 methods, 10^6 reagents, 10^2 solutions, 10^3 conditions.

game, are players driven by electricity more likely to discover targets faster and more often than players driven by chemical energy? This question is actually difficult to answer because both intelligence and physical strength are indispensable in a treasure hunt game. Electricity may make physical activities easier, but this logic does not necessarily apply to intelligence. So, let's switch to a different analogy: if the same person drives a Tesla and rides a horse, which situation is more likely to discover targets faster and more often? The answer to this question is probably self-evident.

So, can using intelligent machines alone accelerate the discovery of new materials and processes? The answer is also negative. Machines only provide a linear increase in experimental operations compared to human researchers, and this linear coefficient is not particularly large. From my personal experience, it may be around 20 times. The biggest problem with linear relationships is their linear relationship with time, meaning that achieving large-scale knowledge innovation will take a very long time in the future. This is not a reasonable scientific framework; it lacks imagination. Let's imagine a completely new mathematical formula:

$$F_{n+1} = F_n + F_{n-1}$$

Now let's define $n+1$ as tomorrow, n as today, and $n-1$ as yesterday. Let the function F represent the amount of data in a particular industry or chemical experiment. Then this sequence is the famous Fibonacci sequence, with the first few terms being: 1, 1, 2, 3, 5, 8, 13, 21, 34, and so on. The general form is: $F_n = \frac{\phi^n - (-\phi)^{-n}}{\sqrt{5}}$, where $\phi = \frac{1+\sqrt{5}}{2}$. When n is large, F_n is proportional to ϕ^n. Finally, we see an exponential growth relationship.

At first glance, does this look very familiar? Yes, it is indeed the mathematical principle of the famous golden ratio, where the artistry of science is vividly demonstrated. Taking a closer look, it becomes even more familiar. Essentially, this is what current large language models are doing: packaging all historical data for training, starting with F_{n-1}, and then incorporating all incremental data from today onward. For text, video, and other information on the internet, due to relatively uniform data formats and the accumulation of technical expertise in this area, technologies like ChatGPT and Sora have been impressive since their inception. If it were possible to generate new chemical materials with just a prompt word like Sora, how would we need to reconstruct our databases and algorithms?

References

Arrhenius, S. (1889). Über die Dissociationswärme und den Einfluss der Temperatur auf den Dissociationsgrad der Elektrolyte. *Zeitschrift für Physikalische Chemie* 4 (1): 96–116.

Ashcroft, N.W. and Mermin, N. (1976). *Solid State Physics*. Saunders Collage Publishing.

ChinaEducationOnline (2015). The popularity of postgraduate applications slows down and the enrollment crisis begins to spread. Retrieved from https://www.eol.cn/html/ky/report2015/a2.shtml.

Griffiths, D.J. and Schroeter, D.F. (2018). *Introduction to Quantum Mechanics.* Cambridge University Press.

Miyazaki, A. (2023). New Nature Index data shows China on top for natural sciences as US leads on health sciences research. Retrieved from https://www.natureasia.com/en/info/press-releases/detail/8966.

Noether, E. (1918). Invariante variationsprobleme. *Nachrichten von der Gesellschaft der Wissenschaften zu Göttingen, Mathematisch-Physikalische Klasse*, pp. 235–257.

Röntgen, W.C. (1896). On a new kind of rays. *Science* 3 (59): 227–231.

Shirakawa, H. and Ikeda, S. (1971). Infrared spectra of poly (acetylene). *Polymer Journal* 2 (2): 231–244.

SinaEducation (2010). Old photos of Beijing University of Technology: Chemistry Experiment Class in the Laboratory (pictures). Retrieved from http://edu.sina.com.cn/gaokao/2010-07-13/1131258180.shtml.

VoxChina (2018). China's Overwhelming Contribution to Scientific Publications. Retrieved from https://voxchina.org/show-3-99.html.

XU, Y., Ye, S., Liu, R. et al. (2024). Spider Matrix: Towards Research Paper Evaluation and Innovation for Everyone. *ChemRxiv.* doi: 10.26434/chemrxiv-2024-689vs-v4.

2

Robots Technology Development in Modern Scientific Research

2.1 Introduction

The development of intelligent laboratories is a saga of technological evolution that has reshaped both industry and academia. Spanning over five decades, this transformation has been punctuated by significant milestones in robotics and artificial intelligence (AI), which have fundamentally altered the approach to material, chemical, and biological sciences.

The period from 1978 to 1999 laid the groundwork for laboratory automation with seminal developments such as Merrifield's automated peptide synthesis, which not only automated the tedious process of peptide chain assembly but also introduced the concept of programmable machines in laboratory settings. During this time, advances in robotics were primarily centered around manufacturing, but the principles of automation and control systems began seeping into laboratory environments, setting the stage for future integration of more advanced computational methods and AI.

From 2000 to 2010, the early robotic systems were refined and made more adaptable to a variety of laboratory tasks, leading to an uptick in their academic and industrial applications. Key developments during this period include the establishment of robotic arms for sample handling and the introduction of machine learning algorithms for data analysis, which enhanced the capacity for high-throughput screening and compound analysis. Such advancements represented a significant step toward the current state of intelligent laboratories, as they began to perform not just repetitive tasks but also complex sequences of operations that required a certain level of decision making.

The period from 2011 to 2016 witnessed the advent of AI-driven platforms that could handle not only data analysis but also the planning and execution of experimental procedures. Robotics technology saw substantial enhancement in precision and versatility with the introduction of 3D printing technology in laboratories, allowing for the customization of lab equipment and the expansion of automation into the realms of experimental design and synthesis.

Between 2017 and 2019, the emergence of self-driving laboratories exemplified by the work of MacLeod et al. marked a new horizon for intelligent labs. These systems integrated AI to a degree that enabled autonomous hypothesis generation,

AI and Robotic Technology in Materials and Chemistry Research, First Edition. Xi Zhu.
© 2025 WILEY-VCH GmbH. Published 2025 by WILEY-VCH GmbH.

experimental design, execution, and even interpretation of results. The synergy of AI and robotics was no longer a facilitator but rather a driver of scientific discovery, demonstrating the capability to navigate the vast chemical space and optimize synthetic pathways with minimal human intervention.

The most recent phase (2020–2023) has seen a dramatic acceleration in the adoption and sophistication of intelligent laboratory systems, arguably catalyzed by the onset of large-scale AI models. These AI models brought unprecedented capabilities to intelligent labs, such as advanced pattern recognition, predictive analytics, and the autonomous generation of new scientific knowledge. By 2023, intelligent laboratories were not only experimental executors but also originators of new research directions, leveraging deep learning and neural networks to uncover insights beyond human intuition.

These advancements have had a profound impact on both the pace and the nature of scientific inquiry. In academia, intelligent laboratories have expanded the scope of research by enabling complex, multi-step experiments that would be impractical for human researchers to carry out manually. In the industry, the implications have been equally transformative, with intelligent systems accelerating the route from discovery to commercialization, optimizing processes, and enhancing the efficacy of research and development pipelines.

As we consider the future, the trajectory suggests a continued merger of AI, robotics, and laboratory science. Intelligent laboratories are anticipated to not only act as competent assistants to human scientists but also emerge as autonomous entities capable of independent scientific discovery.

2.2 Early Development of Laboratory Automation (Before 2000)

The journey toward the automation of laboratories, especially in the realms of chemistry and biology, embarked upon a significant transition with the advent of early automation technologies. One of the most pivotal moments in this journey was the development of the first automated peptide synthesizer by Robert Bruce Merrifield in the 1960s. Merrifield's invention, which later earned him the Nobel Prize in Chemistry in 1984, was not merely a standalone achievement; it represented a paradigm shift toward the automation of complex and labor-intensive scientific processes.

2.2.1 Early Automation Technologies

The inception and evolution of early automation technologies in the realm of chemical synthesis have been pivotal in shaping the landscape of modern scientific research. From the groundbreaking work of Merrifield in the 1960s to the sophisticated systems introduced by the end of the twentieth century, the journey of automation has been characterized by a relentless pursuit of efficiency, precision, and innovation. This section delves into the milestones of early automation technologies, illustrating their transformative impact through a series of pioneering systems and solutions.

Merrifield's pioneering automated peptide synthesis system is designed to streamline and accelerate the peptide chain construction process on an insoluble solid support. This system, central to Merrifield's Nobel Prize – winning work, showcases an early yet profound application of automation in chemical synthesis. Stored solvents and reagents are transferred into a glass reaction vessel via an all-glass-and-Teflon system, controlled by metering pumps and motor-driven circular selector valves. The apparatus operates entirely automatically, conducting all necessary chemical reactions and intermediate purification steps within a single vessel, thereby revolutionizing the synthesis of polypeptides.

The introduction of Merrifield's automated peptide synthesis system in the late twentieth century marked a significant milestone in the development of laboratory automation technologies. The system's design was predicated on the notion that utilizing an insoluble solid support for peptide chain synthesis would best facilitate automation. Its operational simplicity – where the supporting resin containing the C-terminal amino acid remains in the reaction chamber throughout the entire synthesis – eliminates the need for manual intervention during the polypeptide chain construction.

The automation process involves precise pumping of the appropriate solvents or reagents into and out of the vessel in a sequential order, as determined by two motor-driven circular selector valves engineered by Nils Jernberg. One valve controls the selection of reagents and solvents required for each addition of an amino acid residue, while the other selects the correct amino acid solution for each synthesis cycle. These cycles are meticulously orchestrated by an attached programmer, ensuring each of the 80 operations in one cycle is executed in the correct sequence.

This ingenious system not only significantly reduced the time and effort required for peptide synthesis but also set a precedent for the automation of complex chemical processes. The ability to complete the synthesis of a nonapeptide chain in 32 hours of continuous operation, without manual attention, underscored the efficiency and effectiveness of Merrifield's automated system. Furthermore, the synthesized peptides retained full biological activity, highlighting the system's capability to maintain high yield and purity levels. Merrifield's work not only accelerated the field of peptide synthesis but also opened new avenues for the study of large polypeptides and potentially proteins, showcasing the transformative potential of automation in scientific research.

Figure 2.1 depicts an innovative solution for temperature control in chemical reactions, featuring an automated cooling bath system designed for precise thermal management. The system employs a substantial cooling bath containing dry ice in a chlorinated solvent, situated on a thermovale (thermal elevator). This setup is meticulously controlled by the internal temperature of the reaction vessel. When the temperature exceeds a predefined threshold, the cooling bath is automatically engaged with the vessel, rapidly reducing the temperature to prevent any exothermic reaction hazards. This efficient cooling mechanism, reacting in less than one second to temperature fluctuations, ensures the safe and consistent execution of

Figure 2.1 Advanced temperature control in chemical reactions using a cooling bath system (Merrifield 1965).

temperature-sensitive reactions, such as nitrations, which are conducted around 0 °C but are relatively yield-insensitive to minor temperature variations.

The challenge of managing exothermic reactions and maintaining precise temperature control is a critical aspect of chemical synthesis, particularly for reactions that can potentially run away or burst if not properly managed. The system illustrated in Figure 2.1 offers a robust solution to this problem, leveraging the significant temperature differential between the reaction environment (approximately 0 °C) and the cooling bath (−80 °C) to provide a powerful and responsive cooling action.

The incorporation of a thermovale enables the automated adjustment of the cooling bath's position in relation to the reaction vessel based on real-time temperature measurements. This dynamic approach to temperature management marks a departure from traditional methods, offering a more responsive and safer means of controlling reaction conditions. The rapid response time of the system, reacting in less than one second, is particularly noteworthy, highlighting the advances in automation and control technologies that make such precision possible.

Despite the approximate nature of the temperature control – which is deemed acceptable for processes like nitrations, known for their relative insensitivity to moderate temperature fluctuations – this system represents a significant advancement in automated chemical synthesis. It underscores the ongoing evolution of laboratory technologies aimed at enhancing safety, efficiency, and outcome predictability in chemical processes.

Figure 2.2 Overview of the automated chemical reactor system (Legrand and Bolla 1985).

This approach not only mitigates the risk associated with exothermic reactions but also exemplifies the broader trend of integrating sophisticated automation solutions into chemical research workflows. By addressing specific challenges such as temperature control with innovative technological solutions, researchers can achieve greater control over complex chemical processes, enhancing both the safety and the quality of their experimental outcomes.

Figure 2.2 provides a detailed schematic overview of a classical automated chemical reactor system designed for versatile chemical synthesis. The reactor features essential components such as a stirrer and a cooling system, complemented by a splitter for solvent reflux or distillation controlled by an electromagnetic valve (not depicted in the figure). The heating mechanism employs an electrically resistive silver coating at the reactor's base, while cooling is achieved by submerging the reactor in a cooling bath elevated by a pneumatic jack. The system automates all material transfers, quantifying most in the process, and supports the introduction of solid and liquid reagents through a two-step process involving an intermediate vessel for weighing before transfer to the reactor.

Reflecting on the early automation technologies in laboratory settings, Figure 2.2 encapsulates the transition toward fully automated chemical processing systems. This reactor system embodies the integration of classical chemical engineering principles with modern automation and control technologies. By automating the transfer and quantification of reagents, as well as the control of reaction conditions, the system facilitates a high degree of precision and reproducibility in chemical synthesis.

The reactor's design highlights the capability to handle a diverse array of reagents: four solid reactants, seven liquid reagents, and four solvents, showcasing the system's adaptability to various chemical processes. The use of peristaltic pumps for the controlled introduction of small liquid volumes over extended periods and the provision for non-quantified introductions for tasks such as neutralization underscore the system's sophistication. These features exemplify how early automation efforts have evolved to meet the complex demands of modern chemical research, paving the way for the sophisticated intelligent laboratories we envision today.

Figure 2.3 illustrates the progression of automated chemical synthesis systems, culminating in the integration of various functional modules by 1999. Starting with basic computer-controlled operations such as dosing, stirring, heating, cooling, and refluxing, the diagram traces the advancement in technology over two decades. The rapid development of microchips, computing technology, and precision mechanics facilitated the integration of complex modules for high-throughput screening, purification, and product characterization into comprehensive automated systems.

The late twentieth century witnessed a transformative period in the automation of chemical synthesis, beginning with the control of single chemical reaction

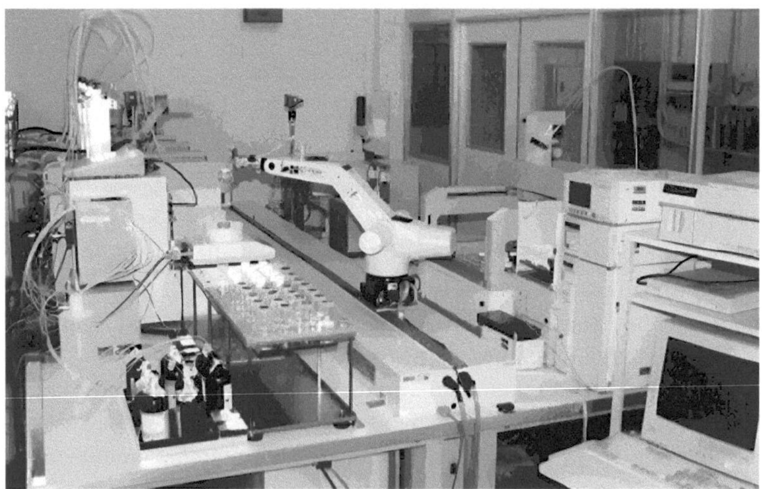

Figure 2.3 Evolution of automated chemical synthesis systems (Okamoto and Deuchi 2000).

processes by computers. These initial systems were capable of automating fundamental laboratory tasks, thereby laying the groundwork for the sophisticated synthesis systems that would follow. The period between the late 1970s and the 1990s was marked by significant technological advancements that would redefine the capabilities of automated systems in chemical research.

As microchip technology, computing power, and precision mechanics evolved, so too did the scope and efficiency of automated chemical synthesis systems. These systems transitioned from performing basic tasks to encompassing an array of functionalities essential for modern chemical research. By 1999, automated synthesis systems had begun to incorporate high-throughput screening modules, enabling researchers to rapidly test a wide array of reaction conditions to identify optimal parameters. This capability significantly accelerated the pace of discovery and optimization in chemical synthesis.

Moreover, the integration of modules for the purification of crude products and the characterization of final products marked a significant leap toward fully autonomous chemical research platforms. These modules allowed for the seamless progression from synthesis to product analysis within the same automated system, streamlining the research process and reducing the time from concept to validation.

The diagram encapsulated in Figure 2.3 not only highlights the technological milestones achieved within this period but also underscores the pivotal role of automation in advancing chemical synthesis. Through the integration of complex functional modules, automated synthesis systems by the end of the twentieth century had become indispensable tools in the chemical research arsenal, capable of undertaking comprehensive research workflows with minimal human intervention. This evolution reflects a broader trend toward increased efficiency, precision, and innovation in scientific research, propelled by advances in automation technology.

Figure 2.4 illustrates the comprehensive solution provided by Chemspeed for automating organic process R&D workflows. Emphasizing a versatile combination of automated batch and flow chemistry, the system incorporates quality by design reactors, process excellence, and user-friendly software. This enables the acceleration, standardization, and digitization of organic synthesis workflows in both batch and continuous modes. The figure's interactive workflow panel demonstrates example workflows facilitated by Chemspeed's technology. Each tool or module within the system is designed to perform specific actions akin to those in manual laboratory workflows, offering a virtually limitless configuration of these tools to suit diverse research needs.

In the realm of early automation technologies, Chemspeed's solution represents a significant leap toward the advanced automation of organic process research and development. By seamlessly integrating batch and flow chemistry with cutting-edge reactor design and comprehensive software management, Chemspeed enables a transformation in how chemical processes are developed, optimized, and scaled. The system's modularity allows for the precise emulation of manual laboratory procedures, thereby not only automating but also enhancing the efficiency and reproducibility of complex organic synthesis workflows.

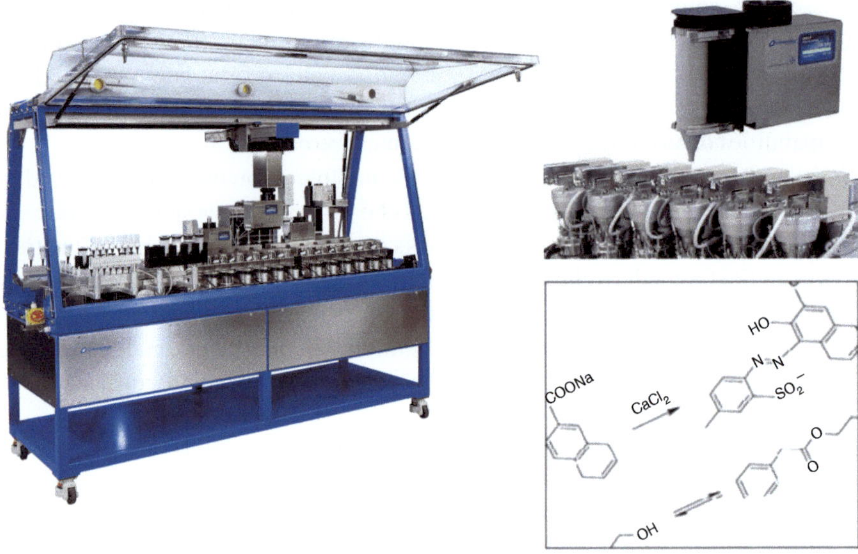

Figure 2.4 Chemspeed's automated batch and flow chemistry solution 1998 (Chemspeed).

Table 2.1 Timeline of early automation technologies in chemical synthesis.

Year	Technology/system	Key features	Impact
1965	Merrifield's automated peptide synthesis system	Automated synthesis on an insoluble solid support	Revolutionized peptide synthesis
1985	Automated chemical reactor system	Automated dosing, stirring, heating/cooling, refluxing	Enhanced safety and consistency in chemical reactions
1998	Chemspeed's automated batch and flow chemistry solution	Integration of batch and flow chemistry with reactor design and software	Accelerated and standardized organic process R&D
1999	Evolution of automated chemical synthesis systems	Integration of high-throughput screening, purification, and characterization modules	Facilitated comprehensive automated synthesis systems

This technology underscores a key trend in the early development of laboratory automation: the move from simple task automation toward the creation of integrated systems that can handle the complexity and variability of scientific experimentation. Chemspeed's approach, with its emphasis on quality by design, process excellence, and user interaction, encapsulates the holistic integration of mechanical automation with digital technologies. As a result, researchers are equipped with tools that not only automate but also innovate, pushing the boundaries of what can be achieved in organic process R&D (Table 2.1).

The trajectory of early automation technologies in chemical synthesis is a testament to the ingenuity and foresight of researchers and engineers who sought to transcend the limitations of manual laboratory processes. Through the implementation of automated systems, from Merrifield's peptide synthesizer to Chemspeed's integrated solutions, the field has witnessed a paradigm shift toward greater efficiency, accuracy, and scope of scientific investigation. These innovations have not only streamlined the synthesis process but also opened new avenues for research and development by enabling the exploration of complex chemical spaces with unprecedented speed and precision. As we reflect on these developments, it becomes clear that the early endeavors in automation have laid a solid foundation for the future of automated and intelligent laboratories, promising a new era of scientific discovery and technological advancement.

2.2.2 Laying the Foundation for AI and Robotics

The advent of early automation technologies in chemical synthesis laboratories catalyzed a pivotal transition, ushering in an era where manual labor-intensive tasks began to be systematically replaced by automated processes. However, the true revolution was yet to come, with these initial steps in automation setting the stage for a more profound transformation through the integration of AI and robotics into laboratory workflows. This progression was not just about replacing human labor but also about augmenting the capabilities of the laboratory environment to facilitate more complex, precise, and insightful scientific inquiry.

As automated systems became more prevalent in laboratories, they started generating vast quantities of data. This surge in data production was both an opportunity and a challenge. It necessitated the development and integration of sophisticated data analysis techniques, thus heralding the introduction of computational methods into the laboratory setting. The 1980s saw the establishment of laboratory information management systems (LIMS), an early incarnation of digital integration in the lab. These systems were designed to manage, store, and interpret the deluge of data produced by automated instruments, representing a significant leap toward the digitalization of the laboratory space.

The transition from manual to automated processes and the consequent need to manage large data sets laid the groundwork for the early stages of AI in laboratories. In the 1990s, with the advent of more powerful computing resources, machine learning techniques began to be applied within laboratory settings for tasks such as pattern recognition and predictive modeling. These early applications of AI were rudimentary by today's standards but marked a significant milestone in the evolution of laboratory automation. They extended the capabilities of automated systems from mere replication of manual tasks to more sophisticated functions like the interpretation of experimental results and aiding in decision-making processes.

Parallel to the integration of computational methods, the late twentieth century witnessed significant advancements in robotics, further enhancing laboratory automation. Robotic arms, which were initially deployed for simple repetitive tasks like moving plates or samples, began to incorporate sensors and vision systems.

This integration allowed for more complex operations within the lab, such as the precise identification of samples and the adjustment of experimental parameters in real-time based on feedback mechanisms. The sophistication of these robotic systems transformed them from basic task executors to integral components of a dynamic laboratory ecosystem, capable of performing complex manipulations and adjustments autonomously.

The confluence of AI, robotics, and automated systems heralded a new dawn for scientific research, characterized by increased efficiency, accuracy, and the capability to perform complex experiments with minimal human intervention. The integration of AI and robotics into laboratory workflows didn't just automate tasks; it fundamentally transformed the nature of experimentation. Laboratories equipped with these technologies could undertake experiments that were previously unthinkable, pushing the boundaries of scientific exploration.

Moreover, the ability of AI and robotics to learn from data and adapt to new information paved the way for laboratories that could optimize their own processes, learn from past experiments, and progressively improve their efficiency and output. This adaptability and learning capability signified a shift from static to dynamic automation systems, where continuous improvement and optimization became possible (Table 2.2).

Table 2.2 Comparison of automation technologies in laboratory systems.

Year	Product/ system	Application	Performance	Intelligence	Automation level
1965	Merrifield's automated peptide synthesis system	Peptide synthesis	Manual intervention minimized; efficient synthesis	Basic; automated sequences and reactions	Initial; automation of specific synthesis steps
1985	Automated chemical reactor system	General chemical synthesis	Enhanced safety and consistency; multiple reactions	Limited; mainly control of physical conditions	Moderate; automated material transfers and conditions control
1999	Evolution of automated chemical synthesis systems	Chemical synthesis with high-throughput screening, purification, and characterization	High throughput; efficient screening and analysis	Advanced; machine learning for pattern recognition	High; integrated modules for comprehensive synthesis tasks
1998	Chemspeed's automated batch and flow chemistry solution	Organic process R&D	Accelerated R&D; standardized processes	High; sophisticated reactor design and software integration	Very high; seamless integration of batch and flow chemistry

2.3 Preliminary Integration and Development of Laboratory Automation (2000–2019)

2.3.1 Automation Technologies (2000–2010)

The dawn of the 21st century heralded a new era in scientific research, marked by significant strides in the integration of AI and robotics within laboratory environments. Spanning from 2000 to 2019, this period witnessed the emergence of groundbreaking systems that not only automated complex tasks but also incorporated advanced AI for planning, execution, and analysis of experiments. The advancements in laboratory automation technologies, as illustrated from Figures 2.5–2.14, exemplify the transformative journey from manual to automated and intelligent laboratory operations.

Transitioning from the foundational groundwork laid in the late twentieth century, the dawn of the new millennium marked a significant leap in laboratory automation technologies. Building upon the early automation systems and the integration of AI and robotics, the period between 2000 and 2010 witnessed the emergence of sophisticated, versatile systems designed to tackle complex scientific challenges with unprecedented precision and efficiency. This era was characterized by the deployment of advanced automated systems such as the Symyx Tools benchtop system for coating analysis, the pioneering robot scientist Adam, and the R-Series modular flow chemistry system. Each of these systems exemplified the new wave of innovation, setting the stage for a transformative impact on research methodologies across various scientific disciplines.

Figure 2.5 showcases the Symyx Tools benchtop system designed for comprehensive analysis of coatings on glass and metal substrates. Central to this system is a

Figure 2.5 Symyx tools benchtop system for coating analysis (NDSUResearchPark).

disposable steel sphere attached to a dual-axis force transducer on an XYZ robotic arm, enabling precise friction, tack, wear, and chemical resistance measurements. The system's versatility is enhanced by its ability to select from over 100 spherical probes for testing, accommodating a wide range of experimental conditions. Additionally, the incorporation of challenge liquids or lubricants via a pipette tip, along with wear measurement facilitated by an optical profilometer and digital camera, exemplifies the system's advanced capabilities in monitoring and analyzing coating durability and performance.

In the evolution of materials science research, the need for precise and automated analysis of coating properties has become increasingly paramount. The Symyx Tools benchtop system represents a significant advancement in this field, providing a versatile and efficient solution for assessing various aspects of coatings applied to different substrates. The system's innovative design allows for the automated delivery of micro titer plate format substrates, enabling unattended operations and significantly enhancing laboratory throughput.

The integration of a disposable steel sphere probe with a dual-axis force transducer offers a unique approach to measuring coating properties such as friction, tack, wear, and chemical resistance. This method ensures high precision in applying specified forces to the coating, allowing for accurate and reproducible measurements. The ability to select the appropriate spherical probe from an extensive library further underscores the system's adaptability to diverse research needs.

For wear measurements, the Symyx Tools system employs an optical profilometer and digital camera to meticulously monitor the groove formation due to the probe's abrasive action. This feature not only provides valuable insights into the coating's durability but also enables a detailed analysis of wear patterns, contributing to a deeper understanding of material properties and performance.

Unlike traditional laboratory systems, Adam, a pioneering robot scientist, stands out for its ability to autonomously design experiments to test hypotheses, utilizing intricate internal cycles for this purpose. Key operations include the selection of yeast strains from a freezer library, inoculation into microtiter plates with rich medium, growth measurement, harvesting of cells, and subsequent inoculation into wells with defined media for further growth curve analysis. Adam comprises an automated freezer, liquid handlers, incubators, plate readers, robot arms, plate slides, centrifuge, plate washer, air filters, and a protective enclosure, along with bar code readers, cameras, environment sensors, PCs, and specialized software, enabling the design and execution of over a thousand experiments daily.

The advent of Adam, the robot scientist, marks a significant milestone in the automation and intellectualization of scientific research. Adam's unique ability to independently design experiments and test hypotheses through complex internal cycles represents a leap forward in laboratory automation, merging the realms of AI with biological research. This autonomous system selects specified yeast strains from a comprehensive library, conducts inoculation, and measures growth curves, among other operations, demonstrating an unparalleled level of automation in executing biological experiments.

Adam's infrastructure is meticulously designed to support its advanced functionalities. It includes an automated freezer for strain storage, liquid handlers capable of controlling numerous fluid channels simultaneously, incubators for optimal growth conditions, and plate readers for precise measurement. The integration of robot arms, plate slides, a centrifuge, and a washer, complemented by air filtration systems and a robust enclosure, ensures the seamless operation of the system. Additionally, Adam is equipped with bar code readers and cameras for tracking and monitoring, alongside environmental sensors and personal computers that run the bespoke software necessary for Adam's operations.

The capability of Adam to autonomously design and initiate a vast number of experiments daily, selecting from thousands of yeast strains and defined-growth media, underscores the system's exceptional capacity for high-throughput experimentation. With the ability to take growth measurements at least once every 30 minutes, Adam records accurate growth curves and associated metadata, providing valuable insights into the experiments conducted. This level of detail and precision in experimentation, achieved through automation and AI, exemplifies the potential of robot scientists like Adam to revolutionize biological research by enhancing efficiency, reproducibility, and scalability.

Figure 2.6 showcases the R-Series, a state-of-the-art modular flow chemistry system renowned for its versatility and widespread adoption in the scientific community. Tailored for researchers seeking unparalleled performance, the system is celebrated for its ease of use and adaptability across a diverse range of reactions and processes. Endorsed by major pharmaceutical companies and cited in over a thousand peer-reviewed publications, the R-Series benefits from a robust design that has been refined over 12 years, with more than 570 units currently operational worldwide, demonstrating its reliability and efficacy in facilitating advanced chemical research.

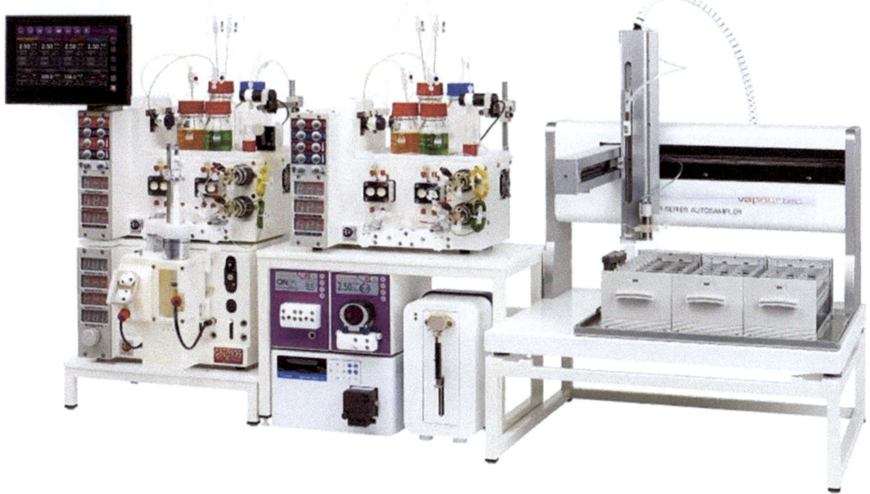

Figure 2.6 R-Series modular flow chemistry system (Vapourtec).

The R-Series stands at the forefront of flow chemistry technology, embodying a culmination of years of innovation and user feedback. Designed to meet the high standards of the scientific community, it offers "best in class" performance, making it a preferred choice for researchers and major pharmaceutical companies alike. Its modular design not only ensures versatility in accommodating various chemical reactions and processes but also simplifies the learning curve, allowing scientists to quickly adapt the system to their specific research needs.

The R-Series' widespread recognition is evidenced by its mention in over a thousand peer-reviewed publications, a testament to its significant impact on the field of chemistry. This peer validation underscores the system's role in advancing research methodologies, contributing to breakthroughs in pharmaceutical development, and other areas of chemical research.

Over 12 years of continuous development and refinement have resulted in a proven design that emphasizes reliability, efficiency, and safety. With over 570 systems currently in use globally, the R-Series' durability and effectiveness in daily operations are well established. Its adoption by leading pharmaceutical companies further validates its status as a cornerstone in modern chemical synthesis, offering scientists the reassurance of a system that is both trusted and innovative.

The R-Series embodies the evolution of flow chemistry systems, providing scientists with a powerful tool for exploring new frontiers in chemical research. Its development reflects a commitment to excellence and a deep understanding of the needs of the research community, ensuring that the R-Series remains at the cutting edge of flow chemistry technology.

Figure 2.7 illustrates the Symyx Core Module robot, a sophisticated system designed for the automated synthesis of new metal–organic frameworks (MOFs). Installed within a glove box under a dinitrogen atmosphere, the robot specializes in automated powder and liquid dispensing into vials with a 1 ml capacity, ensuring meticulous control over the water content in each reaction. Following the

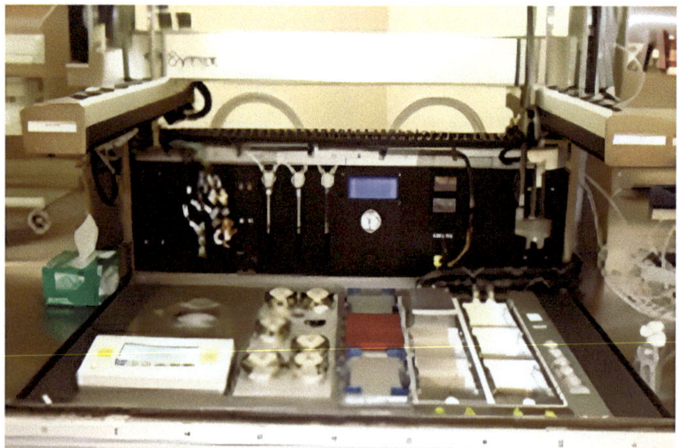

Figure 2.7 Symyx core module robot for the Synthesis of Metal–Organic Frameworks (MOFs) (Haranczyk et al. 2013).

completion of the dispensing routine, the robot facilitates the heating and stirring of individual reaction mixtures directly on its deck. Subsequently, the synthesized products are isolated and prepared for high-throughput X-ray diffraction analysis by transferring them to a glass plate, showcasing the robot's integral role in streamlining the synthesis and analysis of MOFs.

The Symyx Core Module robot represents a pivotal advancement in the field of materials science, particularly in the synthesis of metal–organic frameworks. MOFs, known for their porous structures and potential applications in gas storage, separation, and catalysis, require precise synthetic conditions to achieve desired properties. The automation provided by the Symyx Core Module robot significantly enhances the efficiency and precision of MOF synthesis.

Operating under a controlled dinitrogen atmosphere within a glove box, the robot adeptly manages the challenge of moisture-sensitive reactions by providing precise control over the H_2O content in each reaction vial. This capability is crucial for the synthesis of MOFs, as the presence of water can significantly impact the formation and properties of the resulting frameworks.

The system's ability to automate both powder and liquid dispensing into reaction vials for small-scale reactions is a testament to its versatility and precision. By automating these critical steps, the robot not only reduces the potential for human error but also increases the throughput of experiments, enabling the exploration of a wider parameter space within a shorter timeframe.

Following the dispensing routine, the robot's integrated heating and stirring mechanisms allow for the seamless continuation of the synthesis process. This level of integration ensures that each reaction mixture is subjected to optimal conditions for MOF formation. Finally, the robot's capacity to isolate the products and prepare them for high-throughput X-ray diffraction analysis further exemplifies its role in facilitating rapid synthesis and characterization cycles.

Figure 2.8 illustrates the Emerald Cloud Lab (ECL), an innovative remote laboratory platform that leverages the power of AI and robotics to enable comprehensive scientific research across multiple disciplines from anywhere in the world. The ECL system provides a unique blend of automation and intelligence, allowing researchers to remotely control sophisticated laboratory instruments and conduct experiments with unprecedented precision and efficiency.

Figure 2.8 Emerald Cloud Lab (ECL) (EmeraldCloudLab).

The Emerald Cloud Lab represents a significant advancement in the field of scientific research, offering a fully automated, cloud-based laboratory environment. This system allows researchers to access a wide array of laboratory equipment and perform experiments remotely, removing geographical and physical barriers to cutting-edge research facilities.

ECL's integration of AI and robotics into its platform is at the core of its innovative approach. Through advanced machine learning algorithms, ECL can optimize experimental conditions, predict outcomes, and analyze data with high accuracy. The robotic components of the system ensure precise handling and manipulation of materials, consistent execution of experimental protocols, and maintenance of optimal conditions for each experiment.

One of the standout features of the ECL is its ability to provide researchers with real-time access to their experiments, including live data feeds and analytical results. This immediate access to experimental data not only accelerates the research process but also enhances the ability to make informed decisions on the fly.

Furthermore, the ECL platform emphasizes reproducibility and scalability in scientific research. By standardizing experimental procedures and conditions through automation and AI, ECL facilitates the replication of experiments and the generation of reliable, high-quality data. This focus on reproducibility addresses one of the critical challenges in contemporary scientific research, paving the way for more consistent and verifiable scientific discoveries.

Figure 2.9 introduces the Big Kahuna system by Unchained Labs, a versatile platform designed to transform workflows in the fields of biologics, gene therapy vectors, small molecules, polymers, and catalysts. Offering automated, high-throughput, end-to-end solutions, the Big Kahuna system is fully configurable to meet diverse research needs. It streamlines formulation screening and

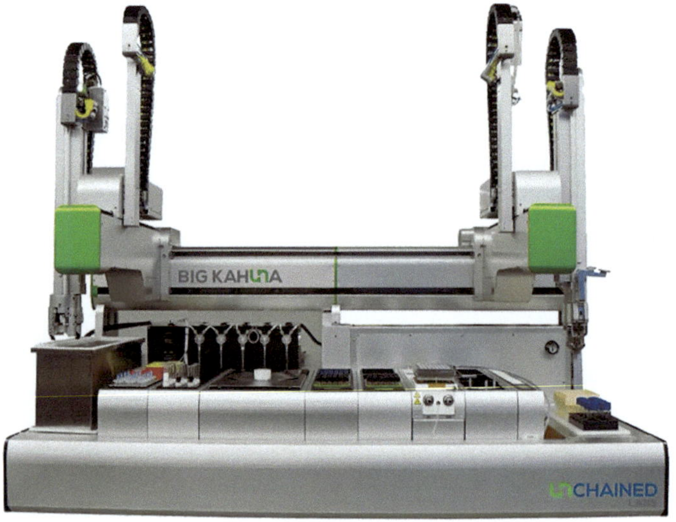

Figure 2.9 Unchained labs' Big Kahuna system (UnchainedLabs).

optimization, visual inspection, sample preparation, reaction screening, powder dispensing, utilization of pressurized reactors, and process chemistry, representing a comprehensive tool for enhancing laboratory productivity and efficiency.

The advent of the Big Kahuna system marks a significant milestone in the automation and optimization of scientific workflows. Developed by Unchained Labs, this innovative platform is engineered to address the multifaceted challenges of modern research laboratories, offering a high degree of automation for a wide range of scientific applications. By facilitating end-to-end solutions for a diverse set of disciplines, the Big Kahuna system exemplifies the latest advancements in laboratory automation technology.

The system's fully configurable nature allows it to adapt to various research requirements, from biologics and gene therapy vectors to small molecules and catalysts. This adaptability ensures that researchers can tailor the platform to their specific workflow needs, maximizing productivity and efficiency.

Integrating the Big Kahuna system into the broader narrative of laboratory automation underscores the continuous evolution of AI and robotics in enhancing scientific inquiry. As technologies like the Big Kahuna become increasingly integral to research laboratories, the potential for innovation and discovery in science grows exponentially. The era of enhanced laboratory productivity, powered by systems like the Big Kahuna, heralds a future where the limits of research are defined not by the constraints of manual processes but by the boundless possibilities of automation and technology.

Figure 2.10 illustrates the transformative role of microreactor technology and flow chemistry in fostering green and sustainable synthesis practices. Highlighting the benefits of a reconfigurable platform, this figure underscores the potential

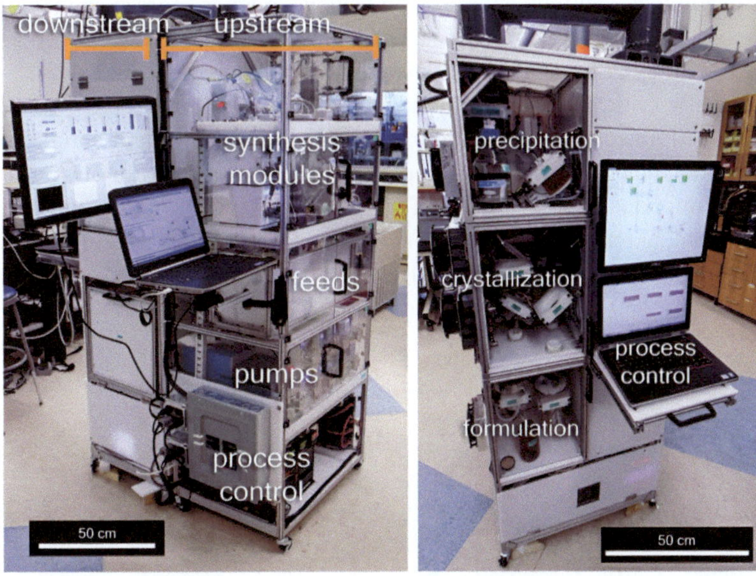

Figure 2.10 Microreactor technology and flow chemistry for green synthesis (Fanelli et al. 2017).

for "on-demand" and "instantaneous" production of short-lived pharmaceuticals. Emphasizing lower production costs, enhanced safety, automation via computer-controlled processes, and significant waste reduction, the system exemplifies an advanced approach to the continuous production and formulation of active pharmaceutical ingredients (APIs).

The integration of microreactor technology and flow chemistry into the production of pharmaceuticals represents a significant shift toward more sustainable and efficient manufacturing processes. This transition not only aligns with the principles of green chemistry but also addresses several challenges faced by the pharmaceutical industry, including the need for cost reduction, safety improvement, and environmental responsibility.

The depicted reconfigurable platform offers a versatile solution capable of adapting to various synthesis requirements, facilitating the "on-demand" and "instantaneous" production of pharmaceuticals. This capability is particularly relevant for short-lived pharmaceuticals, where traditional batch processing methods fall short in terms of efficiency and responsiveness.

As the pharmaceutical industry continues to evolve, the adoption of microreactor technology and flow chemistry is poised to play a pivotal role in developing green and sustainable synthesis methods. This innovative approach not only promises to enhance the efficiency and safety of pharmaceutical production but also reflects a broader commitment to environmental stewardship and resource conservation.

Incorporating such advanced technologies into the synthesis of APIs and other pharmaceutical compounds signifies a forward-thinking strategy aimed at addressing the complex demands of modern healthcare. As depicted in Figure 2.10, the future of pharmaceutical manufacturing lies in leveraging the potential of microreactor technology and flow chemistry to achieve greener, more sustainable, and adaptable production methods, setting a new standard for the industry.

Figure 2.11 presents the innovative automated-flow solid-phase peptide synthesizer, comprising five critical modules. This advanced system automates the peptide synthesis process, from reagent selection and mixing to activation and coupling, ensuring rapid and efficient peptide chain formation. Highlighted within the figure are the system's functionality and the use of an Arduino Mega for prototyping the control system, demonstrating the integration of modern electronics with chemical synthesis for enhanced precision and control.

Figure 2.11 Automated-flow solid-phase peptide synthesizer modules (Mijalis et al. 2017).

The automated-flow solid-phase peptide synthesizer revolutionizes the synthesis of peptides by automating and streamlining the process through a series of specialized modules. Each module is designed to perform a specific function within the peptide synthesis workflow, from the storage and mixing of reagents to the activation of amino acids and the final coupling reaction. This modular approach not only enhances the efficiency of peptide synthesis but also ensures the precision and reproducibility of the process.

The control system for this synthesizer was initially prototyped using an Arduino Mega, showcasing the adaptability and potential for customizability in modern laboratory equipment. The integration of TTL serial ports and the use of standard RS-485 serial protocols for communication with pumps and valves underscore the system's sophisticated electronic control mechanisms. The use of open-source platforms and readily available electronic components, such as the Arduino and SparkFun transceiver breakout, highlights the potential for researchers to adapt and modify the system according to specific research needs.

In sum, the automated-flow solid-phase peptide synthesizer embodies the integration of chemical engineering principles with modern electronics and control theory, marking a significant advancement in the field of synthetic chemistry. This system exemplifies the potential of automation technologies to transform traditional laboratory practices, enabling researchers to achieve more accurate, repeatable, and scalable synthesis processes.

Figure 2.12 unveils an advanced digital chemical synthesis platform that utilizes AI to predict chemical reaction outcomes, retrosynthesis pathways, and experimental procedures. This innovative tool empowers researchers to optimize synthesis methods and automatically generate procedures for both manual and automated laboratory operations. Leveraging molecular transformer models trained on 2.5 million chemical reactions, the platform offers a new paradigm in digital chemical synthesis, combining teachable AI models with a robust scientific computing infrastructure for scalable, efficient, and secure data analysis and model training.

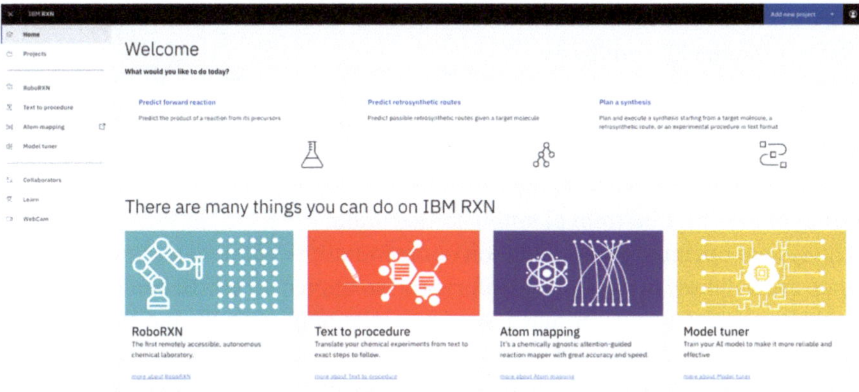

Figure 2.12 IBM RXN for chemistry (IBM).

The digital chemical synthesis platform introduced in Figure 2.12 represents a significant leap forward in the application of AI within the realm of chemical research. By harnessing the power of molecular transformer models that comprehend the natural language of chemistry, this tool transcends traditional rule-based approaches, offering flexibility and adaptability to new data. This breakthrough allows for the accurate prediction of chemical reactions and the generation of optimized synthesis methods, revolutionizing the way researchers approach chemical synthesis.

The platform's underlying technology – molecular transformer models – represents a paradigm shift in digital chemistry. Trained on a vast database of 2.5 million chemical reactions, these models possess an unprecedented understanding of the chemistry's natural language, enabling them to predict chemical behaviors and outcomes with high accuracy. This capability not only accelerates the research and development process but also opens new avenues for innovation in chemical synthesis.

The digital chemical synthesis platform is a part of the broader discovery platform, leveraging advanced scientific computing infrastructure for greater scalability and efficiency. This infrastructure ensures that users can train AI models without limitations on training times or queues, leading to the creation of more accurate models at a faster pace and with better uptime.

The advent of the AI-powered digital chemical synthesis platform, as depicted in Figure 2.12, marks a transformative moment in chemical research. By combining advanced AI predictions with a scalable computing infrastructure, this platform offers researchers a powerful tool for optimizing chemical synthesis, heralding a new era of digital chemistry that is smarter, faster, and more efficient.

An advanced automated synthesis platform is designed for the synthesis of complex organic molecules crucial in the development of small-molecule medicines and functional compounds. This innovative system integrates AI for planning synthetic routes with robotic flow chemistry for execution, marking a significant stride toward fully autonomous chemical synthesis. By learning from millions of previously published reactions, the system suggests synthetic pathways that are then executed by a robotic arm assembling modular process units, demonstrating a novel approach to scalable, reproducible synthesis.

The quest for automated platforms capable of synthesizing complex organic molecules with minimal manual intervention has been a long-standing objective within the scientific community. The platform represents a groundbreaking solution to this challenge, merging the capabilities of AI and robotics to streamline the entire synthesis process from planning to execution.

The system operates on a foundation of computational design, leveraging a comprehensive database of reactions to inform the creation of synthetic routes. Through the application of computer-aided synthesis planning (CASP), the platform can generalize known chemical reactions to new substrates, identify suitable reaction conditions, and predict the likelihood of experimental success. This AI-driven approach significantly reduces the time and effort required for route development and recipe generation, traditionally demanding extensive manual labor from skilled chemists.

Once a synthetic route is proposed, chemists refine chemical recipe files (CRFs) to specify critical parameters such as residence times, stoichiometries, and concentrations compatible with continuous-flow chemistry. The execution phase is managed by a robotic arm that meticulously assembles reactors, separators, and other necessary process units into a continuous-flow system based on the CRF. This robotic orchestration ensures precise control over the synthesis process, enhancing scalability and reproducibility.

The platform's efficacy was demonstrated through the successful synthesis of a suite of 15 medicinally relevant small molecules, showcasing its versatility and its ability to handle varying levels of complexity and stereochemical considerations. This achievement underscores the potential of integrating AI and robotics in chemical synthesis, paving the way for a new era of drug development and materials science.

Reflecting on the period from 2000 to 2019, it is evident that the integration of AI and robotics into laboratory systems heralded a new era of scientific research. The advancements made during this time were not merely incremental improvements but represented a fundamental shift in how research was conducted. These systems, through their innovative designs and applications, demonstrated the potential of automation technologies to not only enhance the efficiency and reliability of scientific experiments but also to unlock new possibilities in research exploration and discovery. As we continue to build on this legacy, the future of laboratory automation, characterized by even more sophisticated AI and robotics integration, promises to further revolutionize scientific inquiry, making the once-impossible within reach.

The evolution of these technologies underscores a pivotal transformation in scientific research methodologies, where the synergy between human ingenuity and machine intelligence continues to push the boundaries of discovery and innovation.

Reflecting on the advancements from 2000 to 2019, it's evident that the integration of AI and robotics has fundamentally transformed laboratory environments. These technologies have moved beyond the automation of routine tasks to enable complex decision-making processes, optimize synthesis methods, and even plan experimental workflows. The period marked a transition from the initial steps of automating physical tasks to a new era where laboratories could operate with a level of autonomy and intelligence previously unimaginable.

As we look toward the future, the groundwork laid during this period sets the stage for further innovations in AI and robotics within scientific research. The continuous evolution of these technologies promises to further enhance the capabilities of laboratories, making research more efficient, accurate, and innovative. The legacy of this era underscores the immense potential of combining human ingenuity with machine intelligence, heralding a future where the boundaries of scientific exploration continue to expand.

2.3.2 Various Forms of Exploration Based on Established Foundations

The onset of the 21st century heralded a transformative era in laboratory automation, underscored by remarkable strides in AI and robotics. This pivotal phase, extending from 2000 to 2019, was marked by the debut of innovative systems that seamlessly melded sophisticated AI with state-of-the-art robotics, thereby significantly broadening the efficiency and scope of scientific inquiry. The precision offered by the **Symyx Tools Benchtop System** in coating analysis, the autonomous experimental prowess of **Robot Scientist Adam**, the versatility of the **R-Series Modular Flow Chemistry System**, and the continued evolution through platforms like the **Emerald Cloud Lab** and the **AI-Powered Digital Chemical Synthesis Platform** underscored a significant leap in scientific process automation.

As we look back on the advancements from 2000 to 2019, it becomes clear that the integration of AI and robotics within laboratory settings has fundamentally redefined the landscape of scientific research. The progress in automation technologies during this period not only heightened the accuracy, safety, and efficiency of scientific endeavors but also ushered in a level of intellectual and operational independence that was once beyond our imagination. The journey through this era – from the initial steps of task automation to the development of fully autonomous systems capable of complex decision making and experimental design – highlights the transformative impact of these technologies.

The **Advanced Automated Synthesis Platform** and systems like the **AI-Driven Automated Microscopy System** exemplify the culmination of these advancements, demonstrating how AI can predict, plan, and execute chemical syntheses, bridging the gap between digital and physical laboratory operations. Moreover, platforms such as the **Digital Chemical Synthesis Platform** and the **Big Kahuna System** by Unchained Labs illustrate the seamless integration of AI for enhancing scalability, reproducibility, and accessibility in research.

Reflecting on this period, the foundation laid for AI and robotics in laboratory environments has not only transformed scientific research methodologies but also set a new benchmark for innovation and productivity in science. As we continue to leverage these advancements, the integration of AI and robotics into scientific research remains a driving force, pushing the boundaries of knowledge and discovery at an unprecedented pace. The legacy of this era stands as a testament to the infinite potential of merging human creativity with machine intelligence, paving the way for future scientific breakthroughs and technological advancements. The era from 2000 to 2019 will undoubtedly be remembered as a defining moment in the ongoing narrative of scientific evolution, heralding a future where the possibilities are as limitless as our collective imagination (Table 2.3).

2.4 Latest Developments and Current Trends (2020–2023)

2.4.1 Automation Technologies in Five Years

Building upon the foundational advancements in AI and robotics integration detailed in Section 2.3.2, the subsequent phase, spanning the late 2010s into the present, has been characterized by mature industrialization and deeper exploratory

Table 2.3 Comprehensive comparison of AI and robotics integration in laboratory systems.

Year	System/ product	Application	Performance	Intelligence	Automation level
2003	Symyx Tools Benchtop System (Figure 2.5)	Coating analysis on glass and metal substrates	High precision in measurement and analysis	Automated control of testing procedures	High; automated dispensing, temperature control, and measure-ment
2009	Robot scientist Adam	Autonomous experiment design and execution in yeast research	Autonomous design of over a thousand experiments daily	Advanced AI for experiment design, execution, and learning	Very high; fully autonomous experiment design, execution, and analysis
2010	R-Series modular flow chemistry system (Figure 2.6)	Versatile modular system for flow chemistry	Best in class performance, adaptable to various reactions	Intelligent modular design for versatile experiments	High; automated modular system for flow chemistry
2011	Next-generation genomic sequencing platform (Figure 2.7)	Automation of genomic sequencing from sample preparation to analysis	High-throughput sequencing with real-time data interpretation	Advanced AI for predictive analytics and real-time interpretation	Very high; fully automated genomic sequencing and data analysis
2014	Emerald Cloud Lab (ECL) (Figure 2.8)	Remote laboratory platform for a wide range of experiments	Enables remote control and execution of experiments	Cloud-based integration and AI for experiment planning	Very high; remote automation of laboratory operations
2015	Big Kahuna System by Unchained Labs (Figure 2.9)	High-throughput solutions for biologics, gene therapy vectors, etc.	Streamlines formulation screening, sample preparation, etc.	Configurable platform for diverse scientific applications	High; automated end-to-end solutions for varied workflows
2016	Digital chemical synthesis platform (Figure 2.10)	Prediction and optimization of chemical synthesis methods	Accurate predictions of chemical reactions and procedures	AI models trained on millions of chemical reactions	High; automation of digital chemical synthesis and analysis

(continued)

Table 2.3 (Continued)

Year	System/product	Application	Performance	Intelligence	Automation level
2017	Automated-flow solid-phase peptide synthesizer (Figure 2.11)	Peptide synthesis	Rapid synthesis with precise control over conditions	Uses AI for route planning and robotics for execution	Very high; robotic assembly of process units for synthesis
2017	AI-powered digital chemical synthesis platform (Figure 2.12)	Digital chemical synthesis, retrosynthesis, and procedure generation	Scalable AI predictions with high model training efficiency	Molecular transformer models for flexible, adaptive learning	Very high; AI predictions integrated with autonomous laboratory execution
2019	Advanced automated synthesis platform	Synthesis of complex organic molecules for medicines	Streamlines the synthesis process for scalable production	Combines AI for planning with robotics for execution	High; automated synthesis with expert-refined control

efforts in laboratory automation technologies. This era has seen the transition from pioneering automated systems to the widespread adoption and sophisticated enhancement of these technologies across various scientific domains, as illustrated in Figures 2.13 through 2.17.

Figure 2.13 showcases an innovative automated synthesis system designed to revolutionize molecule fabrication by leveraging commercially available hardware and standard flow equipment, all orchestrated via custom software within a compact framework ($1.6\,m \times 1.5\,m \times 0.7\,m$). This system introduces a novel approach to automated small-molecule synthesis, featuring a radial array of continuous-flow modules centered around a switching station for unparalleled versatility and efficiency in chemical processing.

This cutting-edge automated synthesis system removes traditional barriers in organic synthesis, granting seamless access to biopolymers and small molecules through highly reproducible and scalable chemical processes. Unlike previous systems that were limited to either iterative or linear synthesis and required compromises in equipment versatility, this system employs a series of continuous-flow modules for flexible, efficient synthesis without the need for manual reconfiguration between processes.

The system's central switching station allows for precise control over reaction conditions, facilitating sequential and multistep syntheses with variable flow rates, the reuse of reactors for different conditions, and intermediate storage. Its design supports both linear and convergent synthesis approaches, significantly broadening the scope of possible chemical transformations.

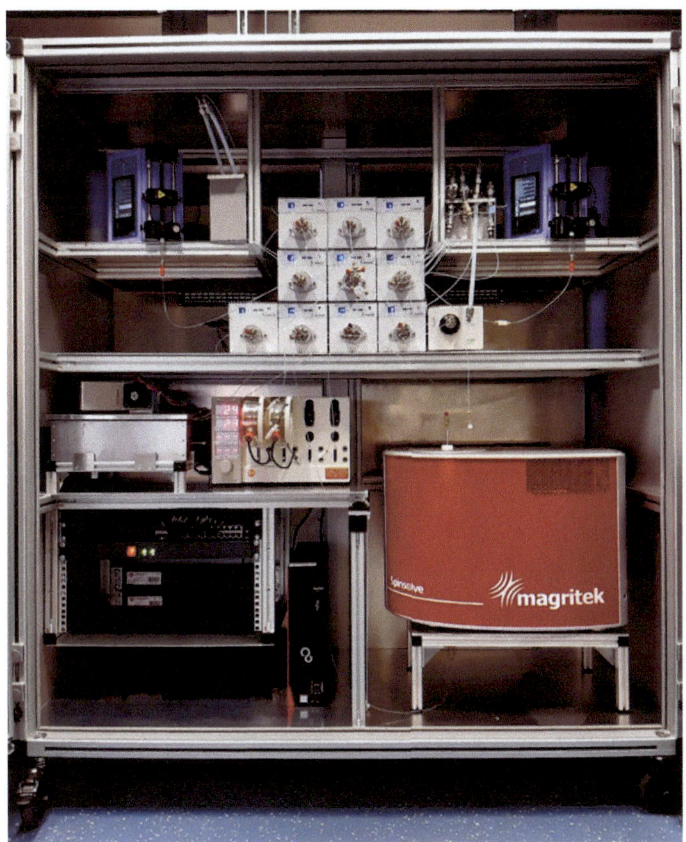

Figure 2.13 Advanced automated synthesis platform for complex organic molecules (Chatterjee et al. 2020).

Demonstrating its robust capabilities, the system has been utilized for a variety of complex synthesis tasks, including the optimization and multistep synthesis of the anticonvulsant drug rufinamide, the creation of derivative libraries through diverse pathways, and conducting metallaphotoredox carbon–nitrogen cross-couplings within a photochemical module. This versatility highlights the system's ability to adapt to various chemical syntheses without needing reconfiguration, marking a significant advancement in automated chemical synthesis technology.

Figure 2.14 illustrates the dynamic integration of robotics in the exploration of photocatalysts for hydrogen production, showcasing the various stages of robotic operation within a laboratory setup designed for high-throughput experimentation.

The integration of a dexterous, free-roaming robot into photocatalyst research represents a monumental shift toward fully autonomous experimental searches in materials science. Operating continuously over eight days, the robot successfully conducted 688 experiments across a ten-variable space, driven by sophisticated batched Bayesian search algorithms. This approach not only optimized the search process but also unveiled photocatalyst mixtures with sixfold higher activity than

Figure 2.14   Advanced robotics in photocatalyst research (Burger et al. 2020).

initial formulations, highlighting the robot's ability to discern and optimize complex material compositions.

The laboratory, equipped with eight distinct stations for comprehensive experiment execution – from photolysis to gas chromatography analysis – illustrates the seamless operation of the robot across multiple tasks and locations, underpinned by advanced navigation and operational capabilities. The robot's mobility and dexterity, enabling it to autonomously load samples, dispense photocatalysts, and analyze results, showcase a novel paradigm where the robot effectively becomes the researcher, navigating and executing within a multi-dimensional experimental framework.

This pioneering application of robotics transcends traditional boundaries of materials research, offering a glimpse into the future where robots could be ubiquitously deployed in various scientific domains beyond photocatalysis. The modular and adaptable nature of this robotic system suggests a broad potential for enhancing research efficiency, precision, and discovery across numerous fields, heralding a new era of innovation in laboratory automation and materials science.

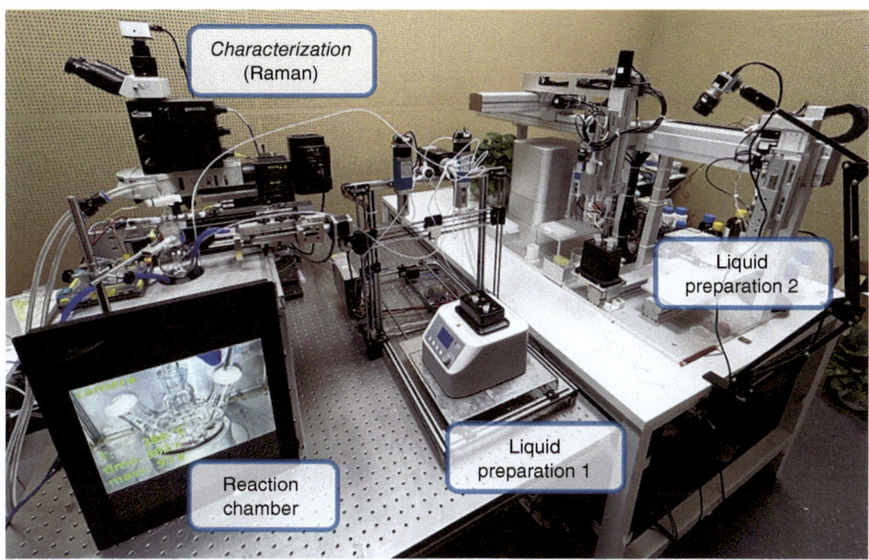

Figure 2.15 Integrating automation with scaling in nanomaterials research (Liu et al. 2022).

Figure 2.15 delves into the innovative blend of automation and scaling within the realm of nanomaterial research, particularly focusing on quantum dots and their chiral properties. (a) Unveils the workflow that marries robotic synthesis with data scaling, powered by neural networks to craft experimental recipes, cycling through this process until specific exit criteria are met, guided by symbolic procedural instructions. (b) Provides an overview of the experimental automation module, highlighting its components like the 10-port valve for liquid reagent distribution, a shaker for ligand exchange, and a reaction chamber tailored for precise temperature and gas environment control. While on-site Raman and absorption spectral analyses are conducted automatically, more complex characterizations necessitating stringent temperature conditions remain manual.

This sophisticated approach to nanomaterial research underscores a significant leap in how experiments are conceptualized, executed, and analyzed. By integrating automated synthesis with advanced scaling methodologies, this system enables a profound exploration of quantum dots, materials known for their potent size effects and chiral properties. The introduction of the chiral dielectric theory, predicated on the exciton absorption mechanism, offers new insights into how chirality influences the dielectric constant via dimensionality, marking a substantial advancement in the understanding of nanomaterials.

The automated intelligent machine (AIM) protocol stands at the core of this innovative system, revolutionizing the generation and interpretation of experimental data through an analytical lens focused on scaling. Demonstrated through the nuanced interpretation of transient absorption data among chiral quantum dots, the AIM protocol unveils discrepancies in the dielectric constant, propelling forward the development of scaling-relevant theories. Additionally, the

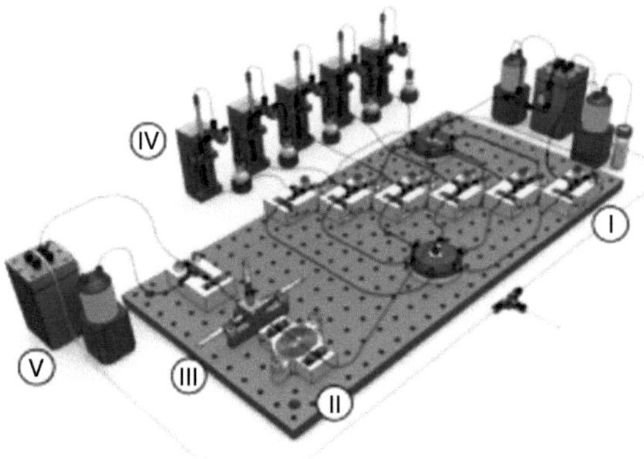

Figure 2.16 AlphaFlow: a self-driven fluidic lab for autonomous discovery (Volk et al. 2023).

protocol's application extends to other spectra analyses, such as absorption and photoluminescence, showcasing its versatility and depth in guiding nanomaterial research.

Figure 2.15 represents a paradigm shift in nanomaterials science, where automation not only accelerates experimental throughput but also enhances the depth of theoretical understanding. The fusion of robotic synthesis with neural network–driven data scaling paves the way for a more nuanced discovery and development process in nanomaterials, embodying the future of research where automation and advanced computational methods converge to uncover new scientific frontiers.

Figure 2.16 introduces AlphaFlow, an advanced self-driven fluidic lab designed for the autonomous exploration and optimization of complex multi-step chemistries. The schematic illustrates the full reactor system, comprising modules for (I) reagent injection, (II) droplet oscillation, (III) optical sampling, (IV) phase separation, (V) waste collection, and (VI) refill, enabling a closed-loop, material-efficient exploration of vast reaction spaces without manual intervention.

AlphaFlow represents a groundbreaking approach to the autonomous exploration of advanced materials and chemistries, leveraging reinforcement learning alongside a modular microdroplet reactor system. This innovative setup facilitates a wide array of reaction sequences, including phase separation, washing, and continuous *in situ* spectral monitoring, thereby addressing the challenges associated with data-sparse environments and complex, multi-step processes.

In a demonstration of its capabilities, AlphaFlow ventured into the realm of semiconductor nanoparticle synthesis, specifically the shell growth of core–shell nanoparticles, inspired by colloidal atomic layer deposition (cALD). Without relying on established cALD parameters, AlphaFlow autonomously identified and optimized a novel reaction route involving up to 40 parameters, achieving superior results compared to traditional sequences.

This achievement underscores the potential of closed-loop systems guided by reinforcement learning to navigate and unravel the intricacies of multi-step nanoparticle syntheses, solely based on data generated within its microfluidic

Figure 2.17 Advanced automated laboratory equipment series by Kapok (Fine-Fanta).

framework. AlphaFlow's success opens new avenues for accelerated knowledge generation and the discovery and optimization of synthetic routes in various multi-step chemistries beyond colloidal atomic layer deposition.

Figure 2.17 presents the Kapok series of advanced automated laboratory equipment, designed with exceptional safety features including fire prevention, explosion-proofing, and leak prevention. This equipment series, including the Kapok PC01 for catalytic reactions, HT01 for hydrothermal synthesis, MC02 for pharmaceutical chemistry, and FS01 for continuous-flow reactions, showcases robust design and materials to ensure stable operation under extreme conditions. Each model is tailored for specific experimental types, boasting high-throughput capabilities and integrated multi-monitoring systems for enhanced operational safety and continuity.

The Kapok automated laboratory equipment series represents a leap forward in ensuring laboratory safety and efficiency in chemical experimentation. Each device within the series is engineered with cutting-edge materials and structural designs to withstand adverse conditions, effectively mitigating risks associated with fire and explosions, and preventing chemical leaks through superior sealing technology. This dedication to safety is complemented by the integration of advanced monitoring systems that continuously assess and manage potential risks, ensuring the safety of both the experimental environment and personnel.

Each model's dimensions are thoughtfully designed to optimize laboratory space while maximizing experimental throughput, embodying Kapok's commitment to advancing scientific research through innovation in safety, functionality, and efficiency. Through the deployment of the Kapok series, laboratories can tackle a wide array of chemical reactions and processes, enhancing the discovery and development of new materials and compounds with unparalleled safety and precision.

As we look toward the future, the continued evolution of laboratory automation technologies promises to drive further industrialization and exploration within the scientific community. The integration of AI and robotics is poised to unlock new frontiers in research, from the development of novel materials and drugs to the discovery of sustainable chemical processes. The legacy of this era of mature industrialization and in-depth exploration will undoubtedly be marked by the accelerated generation of fundamental knowledge, synthetic route discoveries, and optimization, setting the stage for groundbreaking advancements in science and technology.

In conclusion, the journey from foundational AI and robotics integration to mature industrialization and deep exploration reflects a transformative period in laboratory automation. The advancements captured in Figure 2.13 through Figure 2.17 underscore a pivotal shift toward a future where automated systems not only enhance the efficiency and safety of scientific research but also expand the horizons of human knowledge and discovery.

2.4.2 Mature Industrialization as well as In-depth Exploration

Building on the foundational advancements in AI and robotics integration in laboratory environments from 2000 to 2010 discussed in Section 2.3.2, the subsequent years have seen these technologies transition from nascent innovations to pillars of mature industrialization and in-depth scientific exploration. The evolution detailed through Figures 2.13–2.17 in Table 2.4 encapsulates this journey, showcasing the broadening of scope and the deepening of impact that automated systems have achieved in scientific research.

The era post-2010 has been marked by the widespread adoption and refinement of automation technologies across various domains of scientific research. The Advanced Automated Synthesis System, AlphaFlow, and the Kapok Series of equipment embody the industrialization of automation, where robust, high-throughput, and safe laboratory practices have become standard. These systems demonstrate not only the technical feasibility but also the practical utility of automating complex and hazardous chemical processes, thereby enhancing both productivity and safety in research environments.

The integration of comprehensive safety features, as exemplified by the Kapok Series, alongside the ability of systems like AlphaFlow to autonomously navigate complex multi-step chemistries, underscore the maturity of these technologies. They no longer serve as mere auxiliary tools but as integral components of the modern scientific laboratory, capable of performing tasks with minimal human intervention and maximizing material efficiency.

Concurrently, the field has witnessed a significant deepening of exploratory capabilities facilitated by these automated systems. The Autonomous Robotics in Photocatalyst Research and the Integrated Automation with Scaling initiatives represent pioneering efforts to leverage automation not just for efficiency but also for opening new frontiers in research. By employing advanced AI algorithms and modular reactor systems, these platforms have enabled the discovery and optimization of novel materials and processes, such as the efficient photocatalysts and the scaling laws for quantum dots, which could have profound implications for energy, materials science, and beyond.

Table 2.4 Summary of advanced automation systems.

Year	System/product	Application	Performance	Intelligence	Automation level
2020	Advanced automated synthesis system (Figure 2.13)	Automated synthesis of small molecules	High throughput with robust safety features	Custom software for enhanced control	High; fully automated with safety and stability
2020	Autonomous robotics in photocatalyst research (Figure 2.14)	Exploration and optimization of photocatalysts	Autonomous operation over eight days, 688 experiments	Batched Bayesian search algorithm	Very high; fully autonomous robotic operation
2022	Integrated automation with scaling (Figure 2.15)	Scaling and analysis in nanomaterials, specifically quantum dots	Advanced data scaling and generation of experiment recipes	Neural networks for analytical and scaling-oriented data interpretation	High; automated synthesis with in-depth data analysis
2023	AlphaFlow (Figure 2.16)	Autonomous discovery of complex multi-step chemistries	Optimizes synthetic routes for semiconductor nanoparticles	Reinforcement learning for autonomous operation	Very high; self-driven fluidic lab with modular capabilities
2023	Kapok Series (Figure 2.17)	Diverse chemical reactions, including catalysis and hydrothermal synthesis	Enhanced safety and efficiency in high-throughput experiments	Advanced monitoring systems for operational safety	High; automated handling and processing with safety emphasis

The AlphaFlow system (Volk et al. 2023), in particular, illustrates how the fusion of reinforcement learning with fluidic lab capabilities can lead to breakthroughs in nanoparticle synthesis and other high-dimensionality chemistries. Similarly, the strategic use of neural networks in integrated automation with scaling demonstrates the potential of AI to generate fundamental scientific insights from complex experimental data, pushing the boundaries of what can be achieved through automated research.

As we look toward the future, the trajectory set by the advancements in laboratory automation points toward a research landscape increasingly characterized by machines that can think, learn, and discover alongside their human counterparts. The maturity of industrialization in this field is set to further evolve, with automation technologies becoming even more embedded in the fabric of scientific inquiry, driving efficiency, safety, and innovation.

In parallel, the depth of exploration made possible by these technologies continues to expand, promising a future where automated systems not only replicate human tasks but also uncover new knowledge that accelerates our understanding of the natural world. The era of mature industrialization and in-depth exploration, as captured in Figures 2.13 through 2.17, heralds a new phase of scientific advancement where the potential for discovery is limitless, and the fusion of human ingenuity with machine intelligence paves the way for unprecedented breakthroughs in science and technology.

2.5   Outlook on Future Development

The evolution of smart laboratories over the past decades highlights a transformative journey in the integration and application of AI and robotics within scientific research. Initially, the focus was on introducing automation through mechanical arms and automated devices, aiming to enhance efficiency, precision, and safety in routine tasks. This era saw both academia and industry exploring various automation forms, laying the groundwork for future technological integrations (Figure 2.18).

As we moved into the second decade of the twenty-first century, the capabilities and applications of laboratory automation expanded significantly. The integration of AI began to enhance the autonomy and flexibility of robotic systems, marking a shift toward systems capable of performing more complex tasks with minimal human intervention. During this period, the industry leaned toward integrated workstations for high-throughput testing, while academia continued to investigate a wide array of automation forms.

Post-2020, the landscape of laboratory automation witnessed a paradigm shift away from a heavy reliance on mechanical arms in research labs toward the adoption of AI-driven platforms. This transition highlights a move toward developing systems that embody both the physical execution and the intellectual decision-making processes traditionally undertaken by human researchers. Systems like AlphaFlow, which autonomously navigate complex chemistries, and the Kapok series, which emphasize safety and high-throughput capabilities, exemplify the era's hallmark of intelligent experimentation.

The advent of large-scale models has marked a significant milestone in the evolution of AI and its integration into smart laboratories. Since their emergence, these sophisticated AI models have rapidly become a focal point for both academic researchers and industry practitioners, catalyzing a wave of innovation across various fields of scientific research. The significance of large models lies not just in their ability to process and analyze vast datasets with unprecedented efficiency and accuracy, but also in their potential to fundamentally transform the landscape of laboratory automation and experimentation.

Large models, with their deep learning capabilities and extensive knowledge bases, are uniquely positioned to drive advancements in predictive analytics, experimental planning, and the interpretation of complex scientific data. Their application extends beyond mere automation, offering insights and recommendations that can guide research directions, optimize experimental designs, and uncover patterns that would be imperceptible to human researchers. This has led to an increasing

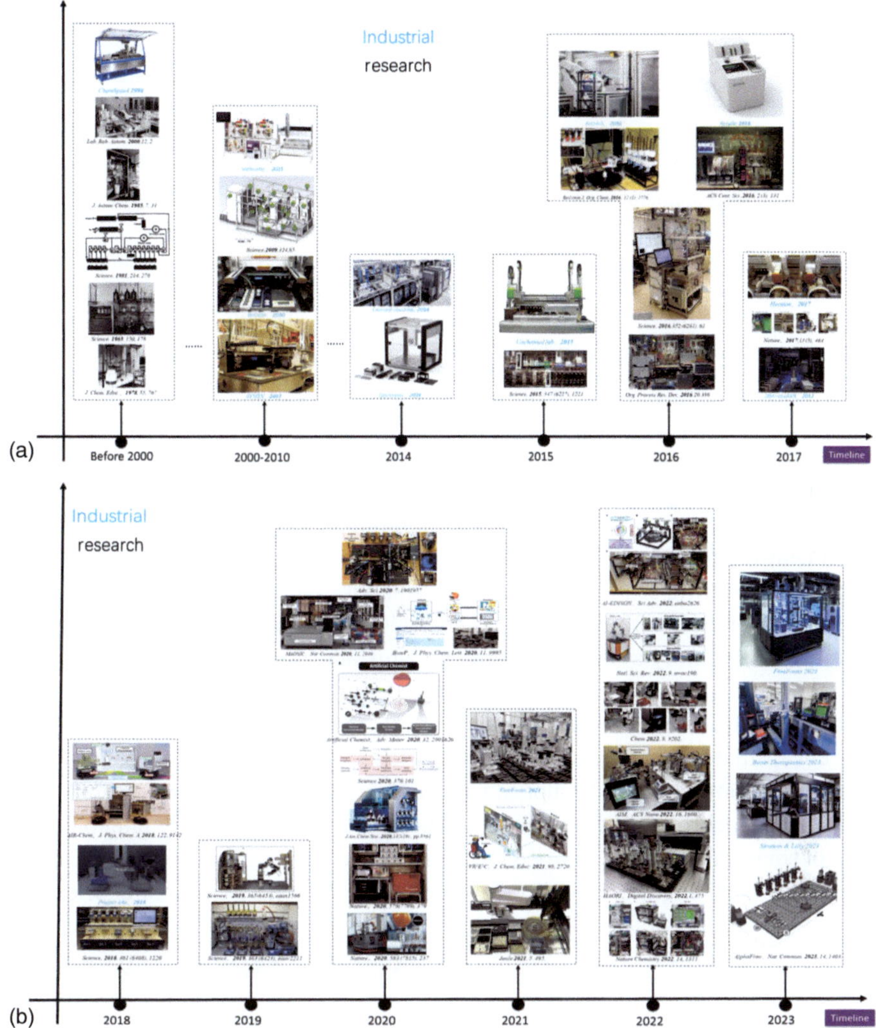

Figure 2.18 Timeline of smart laboratories evolution (Zhu 2023).

number of manufacturers and scholars actively exploring and adopting large-scale models to enhance the capabilities of smart laboratories.

One of the most transformative impacts of large models is their role in accelerating the discovery and development processes within laboratories. By leveraging these AI systems, researchers can navigate the expansive terrain of chemical space, biological interactions, and material properties more swiftly and accurately than ever before. This not only shortens the timeframes for experimental research but also increases the chances of breakthrough discoveries in fields such as drug development, material science, and sustainable chemistry (Table 2.5).

Looking ahead, the future development of smart laboratories is poised for unprecedented growth, driven by advancements in AI and robotics. The trend

Table 2.5 Timeline of smart laboratories evolution and future prospects.

Era	Key characteristics	Technological focus	Impact on research
2000–2010	Introduction of automation	Mechanical arms, automated devices	Enhanced efficiency, precision, safety
2010–2020	Expansion and integration	Integration of AI with robotics	Increased autonomy and flexibility
Post-2020	AI-Driven Smart Laboratories	Advanced AI models, autonomous systems	Deepened exploratory capabilities, operational intelligence
Future prospects	Collaboration and innovation	Large-scale AI models, human–AI collaboration	Accelerated discovery, interdisciplinary research

toward laboratories that are not only automated but truly intelligent – capable of self-directed learning, adaptation, and exploration – suggests a future where AI plays a central role. These systems are expected to feature enhanced AI learning capabilities, fostering seamless human–AI collaboration and allowing researchers to delve deeper into creative and strategic scientific inquiries. Furthermore, the demand for laboratory systems that are customizable, adaptable, and sustainable will continue to grow, reflecting the evolving needs of scientific research.

In essence, the journey from mechanical automation to AI-driven smart laboratories encapsulates the dynamic evolution of scientific research in the digital age. As the integration of AI and robotics within laboratories continues to mature, the promise of unlocking new frontiers in science becomes increasingly tangible. The future of smart laboratories, defined by collaborative systems that not just assist but also actively participate in the quest for knowledge, heralds a new era of scientific discovery and innovation.

2.6 Conclusion

The evolution of smart laboratories, driven by the integration of AI and robotics, has transitioned from foundational mechanical automation to sophisticated AI-driven systems, marking a new era of scientific exploration and discovery. The adoption of large-scale models further amplifies this transformation, enabling unprecedented analytical depth, predictive accuracy, and collaborative potential between human scientists and AI. As we look toward the future, the convergence of these technologies promises to redefine the landscape of research, fostering a more efficient, innovative, and exploratory approach to solving the complex challenges of our time. The journey of smart laboratories reflects not just a technological revolution but also a fundamental shift in the paradigm of scientific inquiry, heralding a future where the potentials of AI and robotics are fully realized in the quest for knowledge and innovation.

References

Burger, B., Maffettone, P.M., Gusev, V.V. et al. (2020). A mobile robotic chemist. *Nature* 583 (7815): 237–241.

Chatterjee, S., Guidi, M., Seeberger, P.H., and Gilmore, K. (2020). Automated radial synthesis of organic molecules. *Nature* 579 (7799): 379–384.

Fanelli, F., Parisi, G., Degennaro, L., and Luisi, R. (2017). Contribution of microreactor technology and flow chemistry to the development of green and sustainable synthesis. *Beilstein Journal of Organic Chemistry* 13 (1): 520–542.

Haranczyk, M., Long, J., Masanet, E., et al. (2013). High-Throughput Discovery of Robust Metal-Organic Frameworks for CO_2 Capture. *Lawrence Berkeley National Lab.(LBNL)*, Berkeley, CA (United States).

Legrand, M. and Bolla, P. (1985). A fully automatic apparatus for chemical reactions on the laboratory scale. *Journal of Analytical Methods in Chemistry* 7: 31–37.

Liu, R., Li, J., Xiao, S. et al. (2022). Authentic intelligent machine for scaling driven discovery: a case for chiral quantum dots. *ACS Nano* 16 (1): 1600–1611.

Merrifield, R. (1965). Automated synthesis of peptides: solid-phase peptide synthesis, a simple and rapid synthetic method, has now been automated. *Science* 150 (3693): 178–185.

Mijalis, A.J., Thomas, D.A. III, Simon, M.D. et al. (2017). A fully automated flow-based approach for accelerated peptide synthesis. *Nature chemical biology* 13 (5): 464–466.

Okamoto, H. and Deuchi, K. (2000). Design of a robotic workstation for automated organic synthesis. *Laboratory Robotics and Automation* 12 (1): 2–11.

Volk, A.A., Epps, R.W., Yonemoto, D.T. et al. (2023). AlphaFlow: autonomous discovery and optimization of multi-step chemistry using a self-driven fluidic lab guided by reinforcement learning. *Nature Communications* 14 (1): 1403.

Zhu, X. (2023). Toward the uniform of chemical theory, simulation, and experiments in metaverse technology. *Precision Chemistry* 1 (4): 192–198.

3

AI Algorithm for Chemical and Bio-material Design

3.1 Introduction

In the field of chemical and bio-material design, the utilization of artificial intelligence (AI) technologies is advancing new research frontiers (David et al. 2020; Mullowney et al. 2023). By digitally representing molecules, AI algorithms can capture the structural and property nuances of compounds at the molecular level, providing material scientists with advanced tools for predicting molecular properties, screening structures, and forecasting reaction mechanisms. This accelerates the discovery process for new materials and supports the development of customized materials effectively.

The rise of data-driven approaches to material discovery has made high-throughput virtual screening a key technique for rapidly identifying compounds with high-performance characteristics. Moreover, the integration of experimental data has significantly enhanced the predictive capabilities of AI models, making the material design process more efficient and accurate. In terms of model interpretability and optimization, researchers are dedicated to developing more transparent and reliable AI algorithms to better understand and guide experimental design.

The application of AI in material science also extends to multi-scale modeling and simulation, offering powerful tools for comprehensive understanding from molecular properties to macroscopic material behaviors. Additionally, AI's role in promoting sustainability and green chemistry practices is evident through its ability to predict and assess the environmental impact of materials, driving the development of more eco-friendly materials (Figure 3.1).

In summary, AI technologies are transforming the field of chemical and bio-material design. Through precise digital representation of molecules and efficient data processing capabilities, they are propelling the entire process from basic research to application development. As technology advances and our understanding deepens, AI is poised to play an even more significant role in shaping the future of material science.

AI and Robotic Technology in Materials and Chemistry Research, First Edition. Xi Zhu.
© 2025 WILEY-VCH GmbH. Published 2025 by WILEY-VCH GmbH.

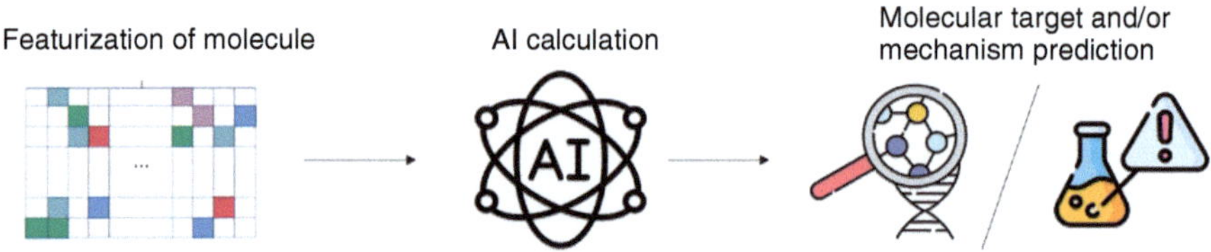

Figure 3.1 General scheme of AI-assistant chemical and bio-material design. The molecules are first embedded into a digital representation, which is further used in AI calculation. Such calculation provides the target molecular structure and/or its mechanism.

3.2 Molecular Representation and Encoding

The fusion of AI with drug discovery necessitates the digital representation of molecules in a format accessible to both scientists and computers. This need has led to the development of various molecular representations (Wiswesser 1968), essential for the computational analysis and visualization of bioactive molecules. Among these, graph-based models for molecules and macromolecules stand out, offering a standardized way to encode complex chemical structures for digital processing.

Key representations such as the International Chemical Identifier (InChI) (Leigh 2011) and the Simplified Molecular Input Line Entry System (SMILES) (Weininger 1988) allow for the detailed yet efficient description of chemical substances. These formats are crucial for AI-driven drug discovery, enabling algorithms to process large compound libraries, predict molecular properties, and streamline the identification of new drug candidates.

In addition to linear notations like SMILES (Weininger 1988), graph representations are gaining prominence in computational chemistry. Graph-based models represent molecules as interconnected nodes (atoms) and edges (bonds), providing a rich structural context that can be leveraged by machine-learning algorithms. These representations capture spatial relationships and connectivity patterns within molecules, facilitating the development of advanced AI techniques for drug discovery.

This section aims to introduce the foundational representations in the field, including linear notations, graph-based models, and representations for chemical reactions, focusing on their application in computational chemistry and bioinformatics. Understanding these representations is vital for researchers at the intersection of computational technology and chemical science, facilitating the exploration of AI applications in drug discovery.

3.2.1 Linear Notations for Molecules

In discussing AI algorithms for chemical and bio-material design, molecular representation and encoding are integral components. To effectively leverage AI technologies, it's crucial to convert the complex structures of compounds into formats that computers can understand and process. This conversion involves various linear notations, which offer concise and information-rich descriptions of small molecules. Among these, SMILES (Weininger 1988), SMILES Arbitrary Target Specification (SMARTS), and InChI are three widely utilized notations. SMILES (Weininger 1988) translates chemical structures into simple strings, SMARTS is used for defining patterns of molecules with specific features, and InChI provides an unambiguous representation of chemical substances. Additionally, beyond SMILES and SMARTS, we will also touch upon other notational systems and encoding strategies that play a pivotal role in molecular representation. These notations not only facilitate efficient storage and retrieval of chemical information but also lay a solid foundation for AI-based compound analysis and design. In the forthcoming content, we will delve into the characteristics, advantages, and applications of these linear notations in the realm of chemical and bio-material design. Through this exploration, we aim to better understand how AI technologies can address challenges in material science and propel the field forward.

SMILES (Weininger 1988): The SMILES notation was conceived as a universal method for expressing chemical structures in a linear, text-based format. Its development stemmed from the need for a standardized chemical language that could be easily comprehended and processed by computers, thus enabling functions such as database organization, chemical analysis, and the advancement of cheminformatics applications. SMILES is built upon its unique capability to unequivocally depict the structure of chemical compounds using concise ASCII strings, making it an invaluable tool for digital chemical communication.

To demonstrate the functionality of SMILES (Weininger et al. 1989), let's examine the example of acetaminophen, a commonly used analgesic and antipyretic medication. In the SMILES representation, acetaminophen can be depicted as CC(=O)Nc1ccc(O)cc1, as illustrated in Figure 3.2a. This notation succinctly conveys the molecule's structure through a depth-first search approach: commencing

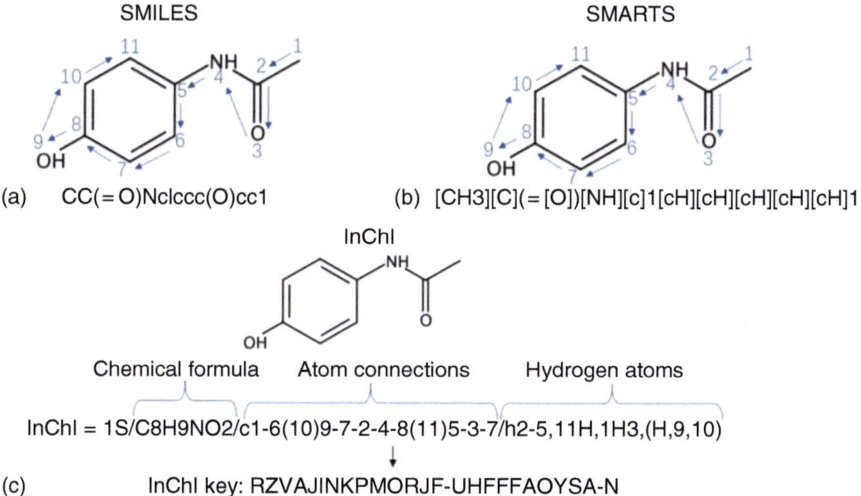

Figure 3.2 The (a) SMILES, (b) SMARTS, and (c) InChI representations of acetaminophen.

with a carbon chain (designated as carbon 1 in Figure 3.2a), introducing a ketone group (C=O), and then detailing a benzene ring containing a hydroxyl group (c1ccc(O)cc1). The brevity of SMILES enables complex molecules to be articulated in a straightforward, linear string, underscoring its efficiency in encoding molecular information.

Nevertheless, it's important to acknowledge that SMILES has its limitations. One challenge lies in the existence of multiple valid SMILES representations for the same molecule, necessitating the use of canonicalization algorithms to derive a unique SMILES notation for indexing purposes. Several algorithms have been developed to tackle this issue; however, the specific algorithm employed can impact the resulting canonical SMILES, emphasizing the need for standardized canonical SMILES generation. Additionally, representing intricate stereochemistry and isotopes, while feasible in SMILES, can introduce complexity into the notation, demanding a deeper familiarity with SMILES conventions to accurately encode and decode molecular structures.

In addition to the existing content, it's worth mentioning that SMILES notation is continually evolving, with ongoing efforts to refine its capabilities and address its limitations. Researchers and chem-informaticians are constantly working to enhance its utility and versatility in the field of chemistry and beyond.

Isomeric SMILES is an extension of the standard SMILES designed to offer a more comprehensive representation of chemical compounds, especially in cases involving isomerism. Isomerism refers to situations where multiple compounds share the same molecular formula but differ in their structural arrangements or spatial orientation. Isomeric SMILES addresses this complexity by integrating additional information into the SMILES notation, including stereochemical details and bond connectivity, to accurately portray various isomeric forms of a molecule. This extension proves invaluable for researchers and chemists when dealing with compounds exhibiting stereoisomerism or other forms of structural variation.

To illustrate the enhancement provided by Isomeric SMILES, let's take the example of 2-butanol, which has two stereoisomers: (R)-2-butanol and (S)-2-butanol. The terms "R" and "S" are derived from the Latin words "rectus" and "sinister," meaning "right" and "left," respectively. They are employed to describe the orientation of substituent groups around a chiral center in a molecule. Specifically, "R" indicates a "rectus" orientation, corresponding to a clockwise arrangement of substituents when viewed from above the chiral center, while "S" signifies a "sinister" orientation, representing a counterclockwise arrangement of substituents when viewed from above. In traditional SMILES notation, both isomers of 2-butanol are depicted by the same SMILES string, "CC(O)C," making it impossible to distinguish between them. In contrast, isomeric SMILES introduces the (R) and (S) prefixes to distinctly represent these two isomers: (R)-CC(O)C (for (R)-2-butanol) and (S)-CC(O)C (for (S)-2-butanol).

This augmentation in isomeric SMILES notation enables precise differentiation between stereoisomers, thereby enhancing the accuracy and clarity of representing complex molecular structures. Such precision proves particularly beneficial in the context of stereochemistry and other specialized areas of chemical research.

Extended SMILES is an advanced variant of the traditional SMILES, designed to provide a more detailed and informative representation of chemical compounds. While standard SMILES notation is excellent for simplifying chemical structures into concise strings of characters, it has limitations when it comes to conveying additional information, such as isotope details, charges, and complex stereochemistry.

Extended SMILES addresses these limitations by allowing for the inclusion of supplemental information within the SMILES notation. This can encompass a wide range of data, such as specifying the isotopic composition of atoms (e.g. using "[2H]" to indicate a deuterium atom), indicating formal charges on atoms or groups (e.g. "[NH4+]" for ammonium), and providing intricate stereochemical descriptions, including cis–trans isomerism and geometric configurations (e.g. "C/C=C/C" for specifying the geometry of a double bond). For instance, consider the molecule 2,3-dichlorobut-2-ene. In traditional SMILES notation, representing the exact stereochemistry can be challenging. However, with extended SMILES, you can accurately describe the molecule's stereochemistry as follows: "Cl/C=C(\Cl)/C=C" to specify the cis arrangement of chlorine atoms on the double bond.

Canonical SMILES is an extension of the SMILES designed to address a fundamental challenge: the existence of multiple valid SMILES representations for the same molecule. Standard SMILES notation can generate multiple representations for a single compound, leading to ambiguity and complicated tasks like database indexing and chemical comparison.

Canonical SMILES overcomes this limitation by providing a systematic and unique representation for each molecule. It achieves this by applying specific algorithms that prioritize and standardize the order of atoms and bonds in the SMILES notation, ensuring a consistent and reproducible depiction. The canonicalization algorithm involves several key steps:

1) **Atom Prioritization**. Atoms are assigned priorities based on their atomic numbers. Atoms with lower atomic numbers are given higher priority. In cases where two or more atoms have the same atomic number, tie-breaking rules are applied.
2) **Bond Prioritization**. Bonds are also prioritized based on their types, with single bonds having lower priority than double bonds, and so on. Like atoms, tie-breaking rules are applied if multiple bonds have the same type.
3) **Traversal Order**. A depth-first traversal of the molecular graph is performed. This means starting from a particular atom and exploring its connected atoms and bonds in a specific order. The order of traversal is determined by the priorities assigned to atoms and bonds.
4) **Branching Decisions**. When branching occurs, such as in ring structures or multiple substituents on a single atom, specific rules are followed to determine the order in which branches are explored.
5) **Canonical SMILES Generation**. As the traversal proceeds, the canonical SMILES notation is constructed step by step, ensuring that atoms and bonds are added in the standardized order dictated by the traversal and prioritization rules.

As a result, regardless of the initial SMILES representation, canonical SMILES consistently represents the same molecular structure. To illustrate the improvement brought by canonical SMILES, consider the molecule "2,3-dichlorobutane." In traditional SMILES notation, two valid representations can be generated due to the presence of identical substituent groups on non-adjacent carbons: ClCC(Cl)CC and CCC(Cl)C(Cl)C. Canonical SMILES, however, resolves this ambiguity by consistently representing the molecule as "ClCC(Cl)CC," ensuring that the same structure is always described using the same canonical SMILES notation.

By applying these rules and systematically traversing the molecular structure, canonical SMILES consistently generates a unique representation for a given molecule. This standardized notation is invaluable in chemical informatics, as it facilitates precise and reliable data management, searching, and comparisons across various chemical databases and applications.

SMARTS (Inc 2020) is a versatile extension of the SMILES notation, as shown in Figure 3.2b. While SMILES primarily represents complete molecular structures, SMARTS is specifically designed to describe chemical patterns and substructures within molecules. SMARTS notation is a powerful tool in cheminformatics and computational chemistry for tasks such as substructure searching, identifying functional groups, and defining molecular features.

One notable improvement that SMARTS brings to chemical notation is its ability to specify complex chemical patterns using a simple and concise syntax. For example, in traditional SMILES notation, representing a specific functional group or substructure within a larger molecule can be challenging. However, SMARTS enables precise pattern matching within molecules.

Here's an example: Let's say you want to find all molecules in a chemical database that contain a phenyl ring. In traditional SMILES, you might need to write lengthy queries. In SMARTS, you can use the pattern "[#6]c1ccccc1" to succinctly describe a phenyl ring, where "[#6]" represents a carbon atom and "c1ccccc1" indicates a six-membered aromatic ring.

International Chemical Identifier (InChI) is a textual identifier for chemical substances designed to provide a standard way to encode molecular information and facilitate the search for such information in databases and on the internet. Developed by the International Union of Pure and Applied Chemistry (IUPAC) (Leigh 2011) in collaboration with the National Institute of Standards and Technology (NIST), InChI serves as a universal, non-proprietary, and structured chemical identifier. Unlike other molecular notations, which are primarily intended for human readability, InChI focuses on being interpretable by software, making it a crucial tool in cheminformatics and digital chemistry.

InChI strings are unique and machine-readable, designed to precisely represent chemical substances without ambiguity. An InChI identifier consists of various layers and sub-layers that encode different pieces of information about a molecule, including its chemical formula, atomic connectivity, and stereochemistry as shown in Figure 3.2c. This allows for a detailed and scalable representation of chemical substances, from simple inorganic compounds to complex biological molecules.

One of the key advantages of InChI is its ability to facilitate interoperability among various chemical databases by providing a consistent identifier for chemical substances. This consistency helps reduce redundancy and confusion in chemical data management and enables more efficient data mining, substance identification, and exchange of information across different platforms and applications. InChI identifiers are widely used in academic research, pharmaceutical discovery, environmental science, and regulatory compliance, among other fields. They are particularly valuable for cataloging compounds in databases, linking related chemical literature, and supporting the development of cheminformatics tools for drug discovery and chemical analysis.

InChI represents a significant advancement in the standardization of chemical identifiers. Its development has greatly enhanced the ability of scientists and researchers to share and search for chemical information globally. Despite its limitations, the widespread adoption of InChI across various chemical information systems underscores its importance and utility in the field of cheminformatics and beyond.

International Chemical Identifier Key (InChIKey), developed by the IUPAC, is a condensed alphanumeric version of the International Chemical Identifier (InChI) designed for easy digital sharing, indexing, and searching of chemical substances. It compresses the detailed InChI string into a short, 27-character code arranged in three blocks, facilitating efficient database queries and internet searches. This fixed-length identifier uniquely represents chemical structures, encoding core molecular information and chemical properties, including stereochemistry, in a format ideal for digital environments. While the InChIKey's compact nature aids in the rapid retrieval and exchange of chemical data across various platforms – essential for sectors like research, pharmaceuticals, and environmental science – it does not allow for the reconstruction of the original InChI, necessitating lookup tables for detailed information access. Despite this limitation, the utility and widespread adoption of InChIKeys underscore their importance in enhancing global access to and interoperability of chemical information.

Table 3.1 provides a concise overview of various versions and extensions of the SMILES) notation, highlighting their key improvements and applications in the field of chemistry and cheminformatics. Each version, from the traditional SMILES to the advanced Canonical SMILES, offers specific enhancements to represent chemical structures and data more effectively. These notations cater to various needs, including stereochemistry, substructure searching, standardization, and rapid molecule identification. Whether simplifying structural input, handling complex molecular information, or ensuring unique identifiers, these extensions play vital roles in the accurate representation and analysis of chemical compounds and their properties.

3.2.2 Graph Representations for Molecules

Molecular graph representation is a powerful tool that maps the structure of molecules to the framework of mathematical graphs. In this approach, atoms become *nodes* and bonds turn into *edges*, making complex chemical structures analyzable through graph theory (West 2001). This connection allows the use of mathematical techniques for chemical problem-solving, such as utilizing algorithms for the shortest path to predict reaction pathways.

Table 3.1 Comparison of linear notations.

Representation	Key improvement
SMILES (Weininger 1988)	Basic linear string representation of molecular structures, simplifying input and storage
Isomeric SMILES (Weininger 1988)	Introduces (R) and (S) prefixes to distinguish stereoisomers, enhancing support for stereochemistry
Canonical SMILES (Weininger et al. 1989)	Implements algorithms to generate unique SMILES, eliminating ambiguity from multiple representations
Extended SMILES	Permits the addition of extra details like isotopes, charges, and complex stereochemistry for a more comprehensive representation
SMARTS (Inc 2020)	Allows for describing chemical patterns and substructures, useful for substructure searching and advanced pattern matching
InChI (Heller et al. 2013)	Provides a unique, algorithm-independent molecular identifier, supporting broader chemical information exchange
InChIKey (Southan 2013)	Derived from InChI, it offers a compact identifier for rapid and unique molecule retrieval and identification

Although a variety of methods exist for depicting molecules in graph form, each is rooted in the core mathematical concept of a graph. In mathematics, a graph serves as a crucial construct for modeling the relationships between entities. A graph, G, is described as a collection of nodes, V, and a collection of edges, E, which link pairs of these nodes. Formally, a graph is denoted as $G = (V, E)$, where each edge is a tuple (v,w) that signifies a linkage between nodes v and w. Graphs are primarily divided into two categories: undirected graphs, in which edges lack orientation, rendering the pair (v,w) identical to (w,v); and directed graphs (or digraphs), where edges are directional, extending from one node to another, thereby distinguishing (v,w) from (w,v). Most molecules are represented by undirected graphs, reflecting the non-directional nature of chemical bonds between atoms (West 2001).

To efficiently represent a graph on a computer, transforming its abstract nodes and edges into linear structures like matrices or arrays is vital. This step is key for illustrating node connections, necessitating an imposed order on nodes for array encoding – a process that aligns with the unordered nature of sets V and E. The core information includes atom connectivity, atom identities, and bond types in the molecule.

Figure 3.3 demonstrates an example of graph representation. The atom connections are typically shown via an adjacency matrix, where a matrix element 1 indicates a bond between nodes (atoms) and 0 refers to no bond. This matrix, also known as a connectivity matrix, generally doesn't specify bond types. Atom and bond identities are captured in node and edge features matrices, respectively. The rows of the node feature matrix correspond to atoms, detailing features like atom type and implicit Hs. Similarly, the rows of the edge feature matrix represent bonds, encoding bond features. Although one-hot encoding is common for these features,

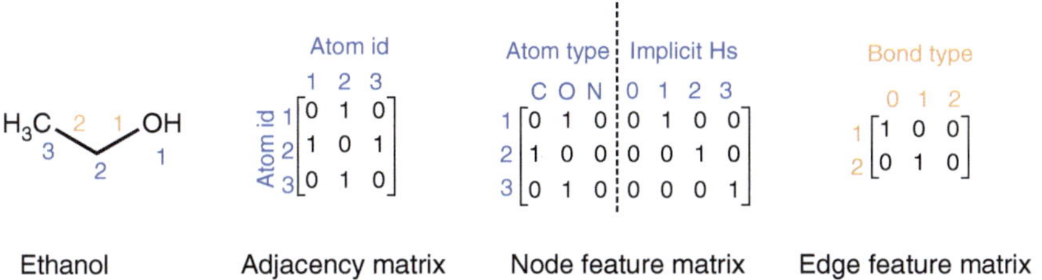

Ethanol   Adjacency matrix   Node feature matrix   Edge feature matrix

Figure 3.3 Example graph representation for ethanol. The node feature matrix can be customized to accommodate specific information requirements. The edge feature matrix can take the form of either one-hot coding or a vector containing relevant data. Implicit Hs represents the number of implicit hydrogen atoms associated with a particular node.

simpler integer-based representations for properties like atom type, charge, and implicit hydrogens are also viable, offering a compact alternative to detailed encoding schemes.

Graph representations for molecules offer significant advantages in molecular design, primarily through their ability to integrate three-dimensional spatial information and enhance interpretability. These representations facilitate a deeper understanding of molecular dynamics, interactions, and functionality by effectively capturing stereochemical details and molecular substructures. The adaptability of graph-based models, particularly graph neural networks (GNNs), allows for nuanced incorporation of structural and chemical properties, making these representations invaluable for a wide array of scientific tasks, from predictive modeling to synthetic design.

However, these representations are not without their limitations. The computational and memory demands of graph representations escalate with the complexity and size of molecules, posing significant challenges in terms of scalability and efficiency. Moreover, the static nature of graph models may inadequately represent dynamic molecular phenomena or complex bonding schemes, such as delocalized or metal–metal bonds. This can lead to inaccuracies or oversimplifications in the portrayal of certain classes of molecules, underscoring the need for careful consideration in their application. Hypergraphs are attempted to handle the problem of multi-valent bonds. In a hyper-graph, instead of tuples of atoms, edges are sets of at least two atoms (hyperedges) (Dietz 1995).

The geometry-enhanced molecular representation (GEM) method (Fang et al. 2022) is designed to encode molecular geometries by simultaneously modeling the relationships between atoms, bonds, and bond angles. It consists of two interconnected graphs: one graph represents atoms as nodes and bonds as edges, while the other graph represents bonds as nodes and bond angles as edges as shown in Figure 3.4. By incorporating these geometric relationships into the network architecture, GEM aims to capture the spatial knowledge of molecules more effectively. This approach enables GEM to learn and utilize the three-dimensional spatial structures of molecules, enhancing its ability to predict molecular properties accurately.

MolGpka (Pan et al. 2021) employs a 1-hot embedding technique for nodes, utilizing a 39-bit scheme to encapsulate various attributes. These attributes include

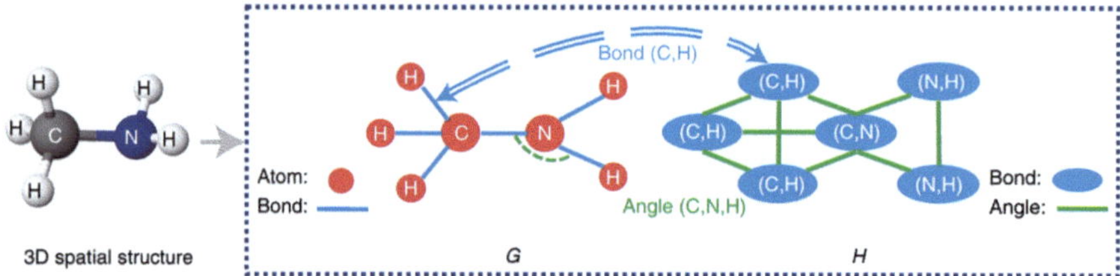

Figure 3.4 Example of geometry-enhanced molecular representation (GEM) (Fang et al. 2022) method. In atom–bond graph *G*, the chemical bonds are regarded as edges, connecting the atoms. In the bond–angle graph *H*, the bond angles are regarded as edges, and a bond angle connects two chemical bonds and three atoms.

the type of atom, its hybridization state, its role as a hydrogen bond donor or acceptor, the atom's connectivity, its valence, its involvement in a ring structure along with the ring's size, its aromaticity, and its status as a protonation site. For edges, the method utilizes a binary adjacency matrix to depict the nodes' interconnections based on the chemical bonds present.

Graph-pK_a (Xiong et al. 2022) similarly employs 1-hot encoded node embeddings akin to MolGpka[13] but with variations in the number of bits allocated for encoding the atom's element, its degree, and hybridization state. Unlike MolGpka, which captures data on the atom's capability to donate and accept hydrogen, as well as its role as a protonation site, Graph-pK_a (Xiong et al. 2022) incorporates details about the atom's formal charge, the presence of radical electrons, the count of bonded hydrogens, and its chirality. Diverging from MolGpka[13], Graph-pK_a (Xiong et al. 2022) also applies a 1-hot encoded embedding vector to edges, regarded as chemical bonds between atoms. This edge-embedding vector conveys information on the bond type, its conjugation status, involvement in cyclic structures, and stereochemical properties.

Epik (Shelley et al. 2007) also employs a node-embedding vector based on chemical information, utilizing a comprehensive 74-bit long vector. This encoding scheme captures various attributes such as the atomic element, atom connectivity, valence and radical electrons count, formal charge, hybridization state, aromaticity, and explicitly bonded hydrogen count. A notable difference in their embedding vector, compared to earlier versions, is the inclusion of a wider range of encoded atomic elements, an increased bit allocation for representing atom degrees, and the incorporation of both 1-hot encoded and continuous values within the embedding. Similar to MolGpka, Epik (Shelley et al. 2007) establishes edges based on the connectivity of bonds among atoms.

3.3 The Formulation of Accessible and Searchable Data

3.3.1 Traditional Way for Molecular Structure–Property Relationship Determination: The Kohn–Sham Equation

The Kohn–Sham equation is a central element of density functional theory (DFT), which is a quantum mechanical method used to investigate the electronic properties of many-body systems, especially atoms, molecules, and solids. This equation provides a practical framework for calculating the ground-state properties of a system of electrons in a non-relativistic quantum mechanical setting. The elegance and efficiency of the Kohn–Sham approach have made DFT one of the most popular and widely used methods in computational physics and chemistry.

The foundation of DFT lies in the Hohenberg–Kohn theorems, which establish that the ground-state properties of a many-electron system are uniquely determined by its electron density. However, these theorems do not provide a practical way to calculate the electron density. The Kohn–Sham equation, introduced by Walter Kohn and Lu Jeu Sham in 1965, addresses this challenge by transforming the many-body problem into a set of non-interacting particles that move in an effective potential.

Mathematically, the Kohn–Sham equations can be expressed as:

$$\left[-\frac{\hbar^2}{2m} \nabla_i^2 + V_{\text{eff}}(r) \right] \psi_i = \epsilon_i \psi_i$$

where $\hbar$ is the reduced Planck's constant, m is the electron mass, ∇^2 is the Laplacian operator representing the kinetic energy of the electrons, V_{eff} is the effective potential experienced by the electrons, ψ_i are the Kohn–Sham orbitals, which are functions that describe the state of each electron, ϵ_i are the eigenvalues associated with each orbital, representing the energy levels of the electrons.

The effective potential V_{eff} is a sum of three terms: the external potential V_{ext} due to the nuclei, the Hartree potential V_{H} representing the electron–electron repulsion, and the exchange-correlation potential V_{xc}, which accounts for the non-classical interactions:

$$V_{\text{eff}}(r) = V_{\text{ext}}(r) + V_{\text{H}}(r) + V_{\text{xc}}(r)$$

The exchange-correlation potential is the most complex component, encompassing all the quantum mechanical interactions among electrons, and its exact form is unknown. Various approximations, such as the Local Density Approximation (LDA) and the Generalized Gradient Approximation (GGA), have been developed to estimate V_{xc}.

In practice, solving the Kohn–Sham equations involves an iterative process as shown in Figure 3.5. An initial guess for the electron density is used to compute

Figure 3.5 General Scheme for the Kohn–Sham equation calculation.

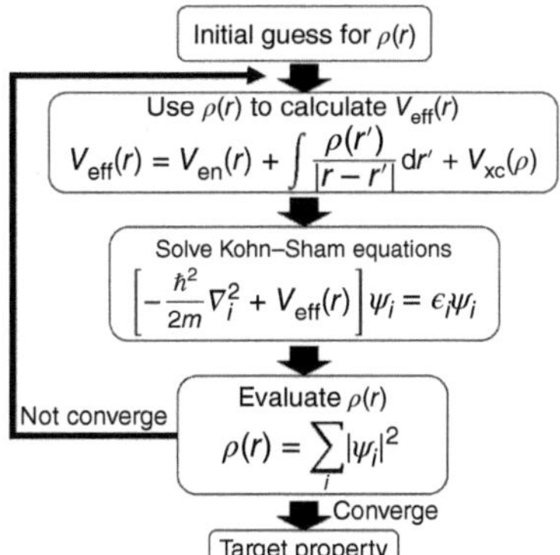

the effective potential, which in turn is used to solve the Kohn–Sham equations for the orbitals and their energies. These orbitals are then used to calculate a new electron density, and the process is repeated until convergence is reached, i.e., until the difference between successive densities falls below a predetermined threshold.

The Kohn–Sham approach has revolutionized the field of computational material science and chemistry by enabling the study of complex systems with remarkable accuracy and computational efficiency. It is applied in the investigation of electronic structure, chemical reactions, material properties, and much more. Despite its approximations, the Kohn–Sham equation provides a powerful balance between accuracy and computational cost, making it possible to explore systems that would be intractable with other quantum mechanical methods.

The Kohn–Sham equation represents a cornerstone in the field of computational quantum mechanics, allowing for the practical application of density functional theory to a vast array of systems. Its development has significantly expanded our ability to understand and predict the behavior of electrons in atoms, molecules, and materials, contributing profoundly to advances in technology, energy, and medicine.

3.3.2 Dataset Preprocessing

The principle of "garbage in, garbage out" is a familiar idea in the realm of machine learning, suggesting that the quality of input data directly affects the output. However, accurately determining a model's predictive capabilities can be difficult without thorough model validation. To effectively evaluate a model, it's crucial to focus on two main aspects: ensuring data is well balanced and conducting model assessments using a separate test set.

Data Balancing. Machine-learning datasets often display heterogeneity, with imbalances manifesting in several ways that can skew model assessments.

The predominance of certain labels is one such imbalance. Take, for example, a binary classification task in drug target interaction, where the dataset comprises 100,000 positive instances against only 1,000 negatives. If this disproportion isn't addressed pre-training, the model might default to predicting a positive interaction for nearly any input, achieving a deceptive 99% accuracy without truly understanding the underlying patterns.

Another common issue is the over-representation of specific features, particularly prevalent in biological datasets. For instance, certain species or molecules might be more thoroughly studied, resulting in data that are biased toward particular sequences or structures. Models trained on such skewed datasets might appear highly accurate, as they perform well for the over-represented groups, but fail to recognize patterns in less common ones. These failures might not significantly impact overall metrics, either because they represent a minor fraction of the test data or because the under-represented groups are entirely absent from the test set. Addressing these imbalances is critical across three stages of model development.

Pre-training Data Preparation. Ensuring a balanced representation involves filtering out duplicates or near-duplicates and proportionally dividing the remaining data across training and testing sets. For sequences, this might mean selecting a single representative from each phylogenetic group; for chemical compounds, clustering by similarity and choosing one sample per cluster can help. This step prevents automatic correct predictions due to (near)-duplicate entries and guarantees that all classes and features are fairly represented in both the training and testing phases.

During Training – Sampling and Weighting. To counteract the model's bias toward predicting the majority class, strategies like adjusting the data weighting or applying oversampling/under-sampling techniques are employed. These methods ensure that less represented classes have a bigger impact on the model's learning process, leading to a more balanced and generalizable performance across different classes and features.

Post-Training Evaluation. It's essential to assess the model's performance in each specific class or feature group independently. This involves evaluating true and false positives/negatives for labels and examining how well the model predicts across different sequence or molecular structure groups. Such detailed evaluation reveals the model's efficacy and fairness, highlighting its strengths and areas needing improvement.

By tackling data imbalance at these stages, it's possible to develop models that are not only accurate on average but also robust and reliable across the spectrum of data they're intended to interpret.

Cross-Validation and Independent Test Sets. Typically, the development of machine-learning models involves iterative adjustments to input features, model parameters, and the choice of the model itself to identify an effective solution. A common pitfall in this iterative process is the repeated use of the same test set for evaluating interim models as well as the final model. This practice compromises the test set's independence, as the iterative adjustments are influenced by performance metrics derived from this test set. Consequently, the risk of the final model being over-fitted to the test set may go undetected. To mitigate this issue, it is imperative to reserve an entirely separate test set prior to the start of model training, using it solely for the final evaluation of the model's performance. During the model development phase, performance monitoring should be conducted using a distinct validation set carved out from the training data or through cross-validation techniques applied to the training dataset. Ideally, multiple iterations should be conducted, with the results' standard deviation reported to statistically verify the significance of any observed improvements. Additionally, when forming validation or cross-validation sets, addressing data imbalance is essential to ensure the reliability of the performance assessment.

3.3.3 Current Existing Dataset

The cornerstone of success in training machine-learning models lies in the utilization of extensive and high-quality training data tailored to each specific molecular property. These invaluable datasets can be sourced from internal, proprietary repositories or extracted from the wealth of publicly available resources. Table 3.2

Table 3.2 Overview of datasets for molecular structure–property relationship used to train AI models.

Dataset	Brief description	Number of samples
DrugBank (Wishart et al. 2008)	Comprehensive drug data including pharmacological, chemical, and pharmaceutical information	15,790
CHEMBL (Gaulton et al. 2012)	Manually curated database of bioactive molecules with drug-like properties	2.4 million
PubChem (Wang et al. 2009)	Database of chemical molecules and their activities against biological assays	115 million
BindingDB (Liu et al. 2007)	Database of measured binding affinities, focusing on protein–ligand interactions	2.7 million
QM9 (Ramakrishnan et al. 2014)	Dataset of molecules with quantum chemical properties calculated at the DFT level	133,885
PDBBind (Wang et al. 2005)	Curated collection of binding affinity data for protein–ligand complexes	23,496
GDB-17 (Ruddigkeit et al. 2012)	Chemical universe database of all molecules up to 17 atoms, used for computational drug discovery	166.4 Billion

provides an overview of popular datasets for molecular properties information used in training many AI applications.

DrugBank (Wishart et al. 2008) is a comprehensive dataset that focuses on registered drugs and their corresponding targets. It contains a wide range of information on 15,790 drugs, with the majority being small molecules and 3392 biologics. In addition to drug–target interactions, DrugBank provides details on the commercial availability status of drugs, their mode of action, and various physicochemical and ADMET properties. This dataset serves as a valuable resource for researchers and practitioners in the field of drug discovery and development, offering insights into the properties and characteristics of a diverse array of pharmaceutical compounds.

CHEMBL (Gaulton et al. 2012) is a prominent dataset that houses bioactive molecular data with drug-like properties, making it a valuable resource for drug discovery research. With information on approximately 2.4 million compounds, CHEMBL offers a vast array of data on various molecular properties and assays, including binding assays, ADME endpoints, toxicology information, and physicochemical properties such as pK_a, solubility, and lipophilicity. Unlike some other datasets, CHEMBL provides SMILES representations of ligands and identifiers for protein targets or references to protein–ligand complexes in the PDB. This rich dataset serves as a crucial tool for researchers seeking to explore the interactions between small molecules and their biological targets, facilitating the development of new therapeutic agents.

PubChem (Wang et al. 2009) is a significant dataset comprising around 115 million compounds and 305 million recorded bioactivity and toxicity data points, making it a valuable resource for researchers in drug discovery and development. This extensive dataset provides a wealth of information on small molecules, including their biological activities and potential toxicities. PubChem serves as a central repository for chemical and biological data, offering researchers access to a wide range of information to support their investigations into the properties and behaviors of various compounds. With its vast collection of compound data and bioactivity profiles, PubChem plays a crucial role in advancing our understanding of chemical compounds and their interactions with biological systems.

BindingDB (Liu et al. 2007) is a comprehensive dataset that contains around 2.7 million binding data points from 1.2 million compounds and 9000 targets, making it a valuable resource for studying protein–ligand interactions. This dataset includes a diverse range of binding affinity data, sourced from both crystallographic structures and docked target series, providing valuable insights into the molecular interactions between proteins and ligands. With a focus on binding affinities measured as constants of dissociation (K_d) or inhibition (K_i), BindingDB offers researchers a wealth of information to explore the binding characteristics of various compounds with their target proteins. By curating a large collection of binding data points, BindingDB facilitates the development of machine-learning models and computational tools for drug discovery and design.

PDBBind (Wang et al. 2005) is a prominent dataset that serves as a valuable resource for studying protein–ligand interactions in drug discovery and structural biology. This dataset contains high-quality crystallographic protein–ligand poses, providing researchers with detailed information on the three-dimensional structures of protein–ligand complexes. By curating binding affinity data for these complexes, PDBBind enables the exploration of binding interactions and the development of predictive models for binding affinity. With a focus on curated subsets of the Protein Data Bank (PDB) database, PDBBind offers researchers access to a diverse collection of protein–ligand complexes with experimental binding affinity values. This dataset plays a crucial role in advancing our understanding of molecular recognition and ligand binding in the context of drug discovery and development.

In summary, the curation of accessible and comprehensive datasets is vital for training precise machine-learning models in the realm of drug discovery. By addressing data challenges and implementing effective strategies, researchers can ensure the robustness and dependability of predictive models for small-molecule properties.

3.4 AI for Molecular Structure–Property Relationship

In this section, the focus is on the models for predicting small-molecule properties. We will first briefly describe the basic deep learning technologies before forwarding to the detailed applications. Moreover, the discussion delves into the advancements in deep learning technology and physical-inspired neural networks for predicting molecular properties. Deep learning models, particularly neural networks, are recognized for their ability to handle complex pattern recognition tasks effectively. The utilization of physical-inspired neural networks, such as graph neural networks, is specifically noted for their capacity to capture intricate connectivity within chemical structures.

Overall, the significance of interpretable machine-learning models is stressed, especially in critical decision-making contexts like drug discovery. While neural networks offer powerful predictive capabilities, their complexity can pose challenges in interpretation. Techniques such as analyzing network weights, and neuron activations, and employing dimensionality reduction methods are suggested to enhance model interpretability. Additionally, future research directions include exploring Bayesian models for uncertainty estimation and enhancing model interpretability to provide effective decision support in drug discovery.

3.4.1 The Deep Learning Technology

A Multilayer Perceptron (MLP) (Delashmit and Manry 2005) is a foundational type of artificial neural network (ANN) (Popescu et al. 2009) composed of an input layer, several hidden layers, and an output layer. Each layer is made up of nodes, or neurons, with the exception of the input layer which directly receives the data. The essence of MLP lies in its ability to learn complex patterns through a process known as feedforward and backpropagation.

In the feedforward process, the input data X is passed through the network. Each neuron in the hidden layers computes a weighted sum of its inputs, $z = \sum_i w_i x_i + b$, where w_i represents the weights, x_i the input values, and b the bias. The result, z, is then transformed by a nonlinear activation function f, yielding the output $y = f(z)$, which is passed on to the next layer.

The training of an MLP occurs through backpropagation, which adjusts the weights to minimize the error between the predicted and actual outputs. This is typically achieved using gradient descent, where the partial derivative of the error with respect to each weight, $\frac{\partial E}{\partial w_i}$ guides the updates to the weights in the direction that reduces the total error.

Through iterative optimization of weights across all layers, the MLP learns to accurately map inputs to outputs, capturing complex nonlinear relationships within the data.

Convolutional Neural Networks (CNNs) (Gu et al. 2018) is a class of deep neural networks, highly effective for processing data with a grid-like topology, such as images. A CNN comprises several layers that automatically and adaptively learn spatial hierarchies of features from input images.

The core building blocks of a CNN are convolutional layers, pooling layers, and fully connected layers. The convolutional layer applies a series of learnable filters to the input. Each filter convolves across the width and height of the input volume, computing the dot product between the filter and local regions of the input, generating a two-dimensional activation map of that filter. Mathematically, for a two-dimensional input X and a filter F of size $K \times K$, the convolution operation (**) is defined as:

$$(F * X)(i,j) = \sum_{m=0}^{K-1} \sum_{n=0}^{K-1} F(m,n) \cdot X(i+m, j+n)$$

where i,j are the spatial indices of the input matrix X. Pooling layers reduce the spatial size of the representation to decrease the number of parameters and computations in the network. The fully connected layers compute the class scores, resulting in the final output predictions.

Training a CNN involves using backpropagation and an optimization algorithm (such as SGD (Amari 1993)) to learn the filter values that minimize a loss function, enabling the network to correctly classify input images. Through its architecture, a CNN can capture the spatial and temporal dependencies in an image through the application of relevant filters, making it exceptionally suited for image recognition and classification tasks.

The landscape of convolutional neural network (CNN) designs has undergone a substantial transformation, with prominent models such as AlexNet (Krizhevsky et al. 2012), VGGNet (Simonyan and Zisserman 2014) (including VGG16 (Qassim et al. 2018), and VGG19 (Mascarenhas and Agarwal 2021)), and ResNet (He et al. 2016) standing out. AlexNet marked a breakthrough with its structure of five convolutional layers—interspersed with max-pooling layers in some cases—and three dense layers. This model introduced innovations like the rectified linear unit (ReLU) activation function, dropout, and data augmentation strategies to boost its accuracy and efficiency.

ResNet (He et al. 2016) represents a further advancement in CNN design, tackling the problem of training exceedingly deep networks by integrating skip connections. These connections promote the direct propagation of gradients, enabling the training of networks with a significantly increased number of layers. This breakthrough has notably enhanced the efficacy of CNNs across a range of image recognition challenges by making the training of deeper networks more practical.

Graph convolutional neural networks (GCNs) (Kipf and Welling 2016) represent a powerful neural network architecture designed for handling data structured as graphs. This approach enables the learning of node representations by efficiently capturing the graph's topological structure. GCNs have gained prominence for their applicability in tasks like node classification, graph classification, and link prediction, where data can naturally be represented as graphs (e.g. social networks and molecular structures).

The foundational operation in GCNs is the graph convolution, which extends the traditional convolution operation from grid-structured data (like images) to graph-structured data as shown in Figure 3.6. The goal is to learn a function that can aggregate features from a node's neighbors and itself to produce a new feature representation.

Mathematically, the layer-wise propagation rule for a GCN can be formulated as:

$$H^{(l+1)} = \sigma\left(\tilde{D}^{-\frac{1}{2}} \tilde{A} \tilde{D}^{-\frac{1}{2}} H^{(l)} W^{(l)}\right)$$

Here, $H^{(l)}$ represents the feature matrix at the l-th layer, where $H^{(0)}$ is the input feature matrix. $\tilde{A} = A + I_N$ is the adjacency matrix of graph A with added self-connections I_N (identity matrix), ensuring that features from the node itself are included in the aggregation process. $\tilde{D}$ is the degree matrix of $\tilde{A}$, and $W^{(l)}$ is the weight matrix for the l-th layer. The normalization factor $\tilde{D}^{-\frac{1}{2}} \tilde{A} \tilde{D}^{-\frac{1}{2}}$ is crucial for

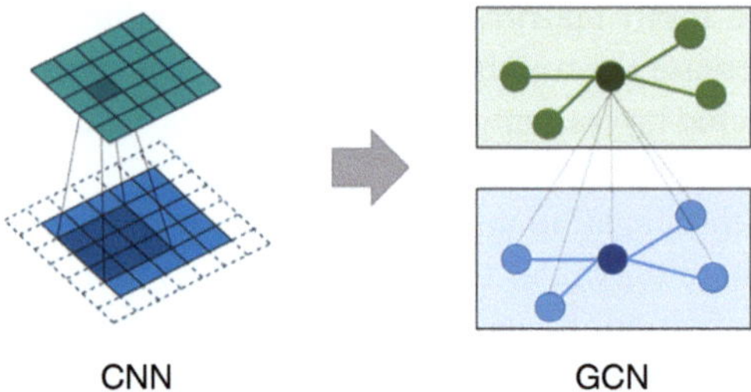

Figure 3.6 Comparison of CNN (Gu et al. 2018) and GCN (Kipf and Welling 2016). CNNs are designed for grid-like data, such as images, where they apply convolutional operations to capture spatial hierarchies. GCNs extend this idea to graph-structured data, applying convolutional concepts to nodes and their neighbors within a graph.

stabilizing the learning process by ensuring that the scale of feature aggregations does not grow with the number of nodes. σ denotes a nonlinear activation function, such as the ReLU function.

Through successive layers, GCNs can integrate and abstract features from increasingly larger neighborhoods, enabling nodes to learn representations that reflect both their local graph topology and their features. This process allows GCNs to effectively model relational data, making them particularly useful for tasks where relationships between entities play a critical role.

The significance of GCNs lies in their flexibility and efficiency in dealing with irregular data, opening new avenues for deep learning applications beyond traditional settings where Euclidean data prevails.

Graph Attention Networks (GATs) (Velickovic et al. 2017) leverage an attention mechanism to dynamically prioritize the influence of neighboring nodes in graph-structured data, enhancing the adaptability and performance of neural network models on tasks like node classification and graph analysis. Unlike traditional graph neural networks that treat connections uniformly, GATs compute attention coefficients to weigh the significance of each neighbor's features, allowing for a more nuanced aggregation of neighbor information.

The core operation in a GAT involves calculating these attention coefficients to determine how much focus should be placed on each neighbor's features. The mathematical formula for computing the attention coefficient between two nodes i and j is given by:

$$\alpha_{ij} = \frac{\exp\left(\text{LeakyReLU}\left(\alpha^T\left[W\,\vec{h_i}\,\|\,W\,\vec{h_j}\right]\right)\right)}{\sum_{k\in N(i)}\exp\left(\text{LeakyReLU}\left(\alpha^T\left[W\,\vec{h_i}\,\|\,W\,\vec{h_k}\right]\right)\right)}$$

Here, W is a weight matrix applied to the node features $\vec{h}$, $\vec{a}$ is a weight vector of the attention mechanism, and $N(i)$ denotes the neighbors of node i. The operation $\|$ denotes concatenation, and α_{ij} represents the normalized attention coefficient computed using the softmax function across all neighbors k of node i. This attention-driven aggregation allows each node to adaptively update its representation by focusing more on important neighbors based on the task at hand.

Through this mechanism, GATs can effectively capture both structural and feature-based information in graph data, offering significant improvements in learning node representations for complex relational tasks without requiring extensive domain knowledge or handcrafted features.

3.4.2 AI Solving the Kohn–Sham Equation

Determination of properties for large chemical substances is essential for applications including batteries, catalysis, and drugs. Machine learning holds the promise of learning energy functional via large-scale samples quickly and accurately, to bypass the time-costing Kohn–Sham (KS) equation solving process (Brockherde et al. 2017). Such a technique has been successfully used to predict molecular dynamics (Zhang et al. 2018) and molecular properties, including the atomization energy (Brockherde et al. 2017; Häse et al. 2017; Hückel 1931; Tsubaki and Mizoguchi 2020), orbital energy (Häse et al. 2017). Pioneer research like graph convolutional network (GCN) (Kearnes et al. 2016), SchNet (Schütt et al. 2018), and deep tensor neural network (DTNN)(Schütt et al. 2017) focus on the final output, considering the atomic coordinates more than their corresponding wavefunctions. Therefore, the transferability cannot be guaranteed for unknown molecules, some of whose features are disparate from the original training set. Such transferability property is important since the generation of large databases is much more time-costing or even impossible (Hansen et al. 2015, 2013; Schütt et al. 2017, 2014) (Figure 3.7).

Message-Passing Neural Networks (MPNN) (Gilmer et al. 2020) offer a flexible framework for learning on graph-structured data, enabling the effective capture

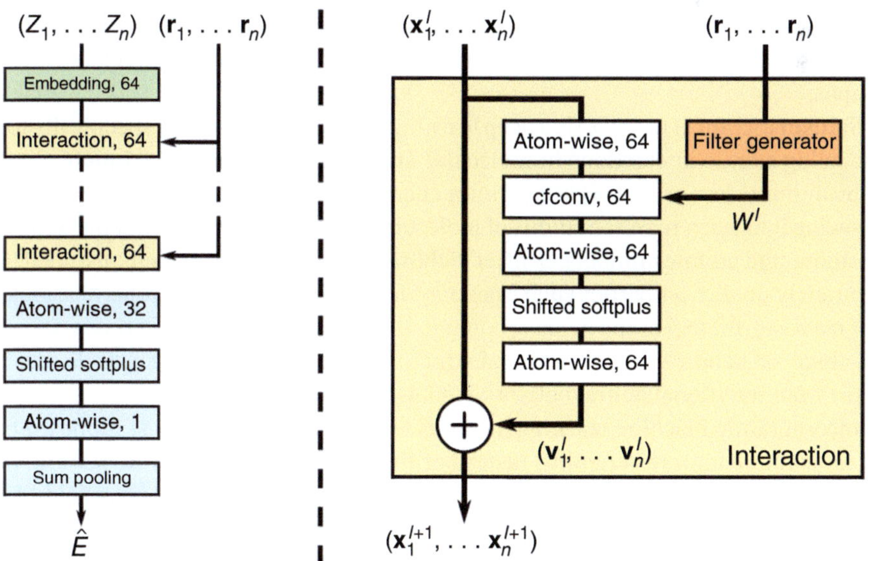

Figure 3.7 Illustrations of the SchNet (Schütt et al. 2018) architecture (left) and interaction blocks (right) with atom embedding in green, interaction blocks in yellow, and property prediction network in blue. For each parameterized layer, the number of neurons is given.

and processing of relational information. Central to the MPNN framework is the concept of message passing, wherein nodes in a graph exchange information with their neighbors over multiple iterations or "steps," allowing for the iterative aggregation and update of node representations based on local neighborhood features.

The MPNN model can be broken down into two key phases: message passing and readout. The message-passing phase involves two functions, M for message computation and U for node state updating, and operates over a fixed number of steps to refine the node representations.

Mathematically, the message-passing phase for a node v can be described as:

$$m_v^{(t+1)} = \sum_{w \in N(v)} M\left(h_v^{(t)}, h_w^{(t)}, e_{vw}\right)$$

$$h_v^{(t+1)} = U\left(h_v^{(t)}, m_v^{(t+1)}\right)$$

Here, $m_v^{(t+1)}$ represents the aggregated message received by node v at step $t+1$, derived from the messages of its neighbors $N(v)$ and potentially the edge features e_{vw}. $h_v^{(t)}$ and $h_w^{(t)}$ denote the states of node v and its neighbor w at step t, respectively. M and U are learnable functions that compute the message and update the node state, respectively.

Following the message-passing phase, the readout phase aggregates the final node representations to produce a graph-level output, which can be used for various downstream tasks such as graph classification or regression.

MPNNs thus elegantly generalize several graph neural network models, allowing for diverse and powerful approaches to learning from graph data by iteratively refining node representations through local neighborhood interactions. This framework has proven effective across a range of applications, from molecular property prediction to social network analysis, by leveraging the inherent structure and features of graphs.

SchNet (Schütt et al. 2018) is a deep learning architecture designed specifically for modeling atomic interactions in molecules and materials. It uses continuous-filter convolutional layers to process quantum chemical properties of atomistic systems, allowing it to learn representations of molecular structures directly from the atomic positions and atomic numbers. SchNet (Schütt et al. 2018) has been demonstrated to accurately predict properties such as energy, forces, and electronic properties, making it a powerful tool in the fields of computational chemistry and materials science. Its ability to handle varying sizes and structures of molecules and materials sets it apart from traditional neural network models.

Incorporating machine-learning methods with underlying physics is supposed to be the solution for transferability. Instead of directly predicting properties from atom positions, AIMNet (Born and Oppenheimer 1927) uses a self-consistent-field-like procedure to iteratively update the initial atomic feature vectors. A similar idea is also implemented in the gradient-domain machine-learning (GDML) approach (Chmiela et al. 2017), where input atoms are mapped into vector-valued functions that obey the energy conservation law. In Gaussian approximation potential (GAP) (Deringer and Csányi 2017; Rowe et al. 2020), properties are obtained

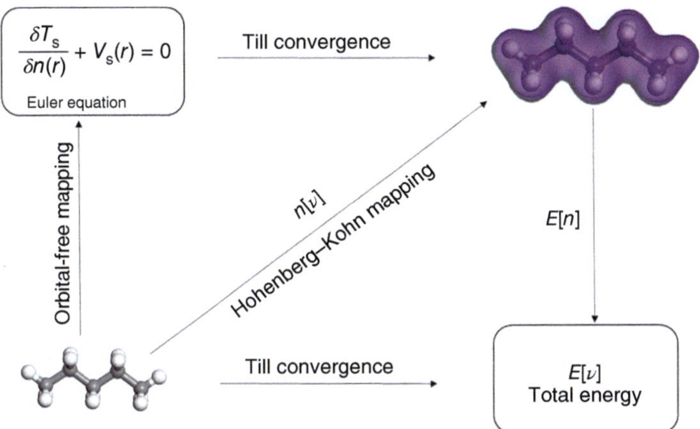

Figure 3.8 The overview of using AI to bypass the Kohn–Sham equation, which is used for structure–property relationship determination. The bottom arrow indicates a conventional electronic structure calculation, i.e., KS-DFT. The ground-state energy is found by solving Kohn–Sham equations. $E[n]$ is the total energy density functional. The center arrow is the HK map from external potential to its ground-state density.

through empirical interatomic potentials as shown in the GAP box in Figure 1a. The non-interpretability of these feature vectors still limits the transferability to unknown material space.

Numerical approximation of wavefunction is another solution to transferability. The Gaussian-type orbital basis is utilized in quantum deep field (QDF) (Tsubaki and Mizoguchi 2020), while the angular information is missing in this isotropic wavefunction. SchNOrb (Schütt et al. 2019) inserts the electronic degrees of freedom by projecting features into fixed directions, which results in a mass of extra parameters, and extends its application for reactive chemistry. In FermiNet (Pfau et al. 2020), the electron wavefunction is constructed in the form of Slater determinants by deep neural networks (DNNs). Without encoding any physical knowledge about wavefunctions besides the essential anti-symmetry, this network is compensated by a much larger amount of optimized parameters. A developed version is PauliNet (Hermann et al. 2020), which replaces the existing ad hoc functional forms (used in the standard Jastrow factor and backflow transformation) with DNNs as shown in Figure 1a, box of PauliNet. The large number of parameters, though have increased PauliNet's accuracy, have also limited its application to systems with less than 30 electrons (Figure 3.8).

Physics-inspired neural networks (Lin et al. 2022; Lin and Zhu 2023; Musil et al. 2021) integrates seamlessly with both data and mathematical models of physics, even when faced with uncertainties, partially understood phenomena, or high-dimensional spaces. This synergy allows for more efficient learning from fewer data samples, making it particularly advantageous in areas where data are scarce or expensive to obtain. Such networks can significantly enhance prediction accuracy and generalization capabilities in scientific and engineering domains by embedding physical laws directly into the learning process.

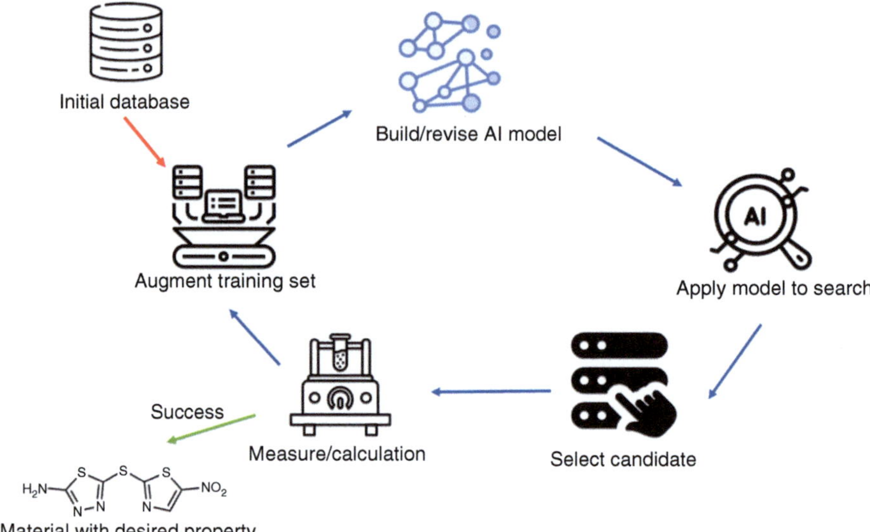

Figure 3.9 Workflow of the AI molecular structure design scheme. Here the measurement or calculation indicates estimation of the material property via first-principle methods or experiments.

3.5 AI for Chemical and Bio-material Design

3.5.1 Design Workflows

Methods based on first principles, machine-learning (ML) predictions, or a combination of both can be applied to the design of polymers. Traditionally, techniques for exploring chemical configuration spaces and designing polymers fall into four main categories:

(I) **Edisonian Design.** This method involves high-throughput computations or experiments to assess the properties of numerous candidate materials. Selection is based on the highest-ranking candidates from these evaluations for further exploration and application, embodying a trial-and-error strategy that relies on random selection or an intuitive grasp of the polymer's configuration and chemical properties.

(II) **ML-Evaluated Design.** Utilizes existing structure–property datasets to develop an ML model, which predicts the properties of various candidates. Top candidates, as determined by ML predictions, undergo further first-principle studies and analyses. Unlike the Edisonian approach, this method uses ML predictions to navigate the material search space, though it primarily excels at interpolating within known property ranges (Patra et al. 2017; Figure 3.9).

(III) **First-Principle-Based Inverse Design.** A promising method that combines optimization algorithms with first-principle calculations or experimental measurements to find optimal parameters for desired properties. This iterative process continues until certain criteria are met, offering a robust approach to polymer design but can be time intensive due to the need for continuous property assessments.

(IV) **ML-Evaluated Inverse Design**. Aims to streamline the first-principle inverse design by using ML models for property prediction, reducing the time spent on direct measurements within design cycles. While promising, its effectiveness hinges on the ML model's ability to predict beyond its training data's scope.

Categories I and II address characterization's forward problem, while III and IV introduce an inverse design strategy, focusing on minimizing the number of evaluated candidates to efficiently identify the target polymer. Both inverse design methods rely on optimization algorithms, with the first-principle approach generating data as needed, and the ML-evaluated method depending on pre-existing data for its predictive model.

While "black-box" optimization algorithms are widely sought after for design and optimization tasks, their effectiveness varies across different problem types due to the "no free lunch" theorems. (Wolpert and Macready 1997) Techniques like Bayesian optimization, genetic algorithms, and Monte Carlo tree search are commonly employed in materials design, with their effectiveness depending on the specific challenges of the task (Chakraborti 2004; Xue et al. 2016; Yang et al. 2017).

3.5.2 Example of Designed Chemical and Bio-materials

The isolation of Pristinin A3 (Yang et al. 2019) was achieved through a meticulously planned purification process. This endeavor began with a large-scale extraction from genetically modified strains of *Streptomyces pristinaespiralis*. Despite initial challenges in securing high yields within large shake flasks, a methodical strategy of dividing the culture into numerous flasks followed by butanol extraction led to the procurement of a crude extract enriched with Pristinin A3 (Yang et al. 2019; Figure 3.10).

The purification process was further refined through the adsorption of this crude extract onto silica gel, utilizing vacuum liquid chromatography (VLC) with gradient elution for initial purification. This step was succeeded by pooling and concentrating the Pristinin A3-rich fractions, which were then subjected to further refinement via preparative HPLC. The apex of this purification journey involved specific column chromatography, employing a solvent gradient for the elution of pure Pristinin A3 (Yang et al. 2019).

To ascertain the identity and elucidate the structure of Pristinin A3 (Yang et al. 2019), sophisticated analytical methods, including LC–MS/MS and NMR spectroscopy, were deployed. LC–MS analysis shed light on the compound's molecular makeup and fragmentation patterns, while NMR spectroscopy revealed intricate structural details, such as specific amino acid residues and post-translational modifications.

The successful isolation and detailed characterization of Pristinin A3 highlight the integration of diverse analytical and purification techniques. This not only enriches the catalogue of RiPP natural products but also unveils novel bioactive compounds with promising therapeutic potentials.

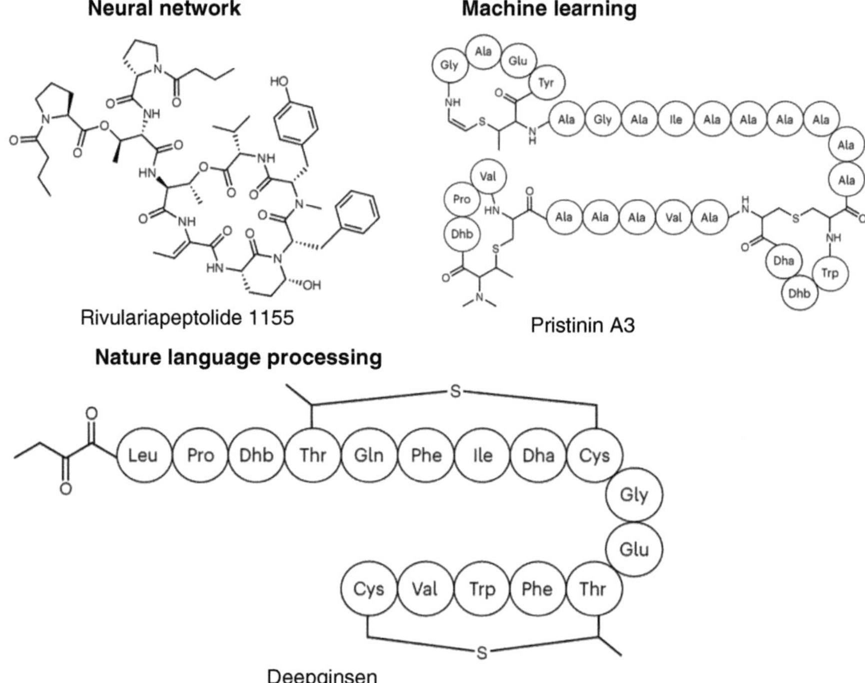

Figure 3.10 Examples of AI-discovered chemical and bio-materials. Rivulariapeptolide 1155 (Yang et al. 2019) is designed by the neural network, Pristinin A3 (Kloosterman et al. 2020) is designed through support vector machine, and deepginsen (Merwin et al. 2020) were discovered in part using natural language processing to predict their RiPP precursors and their cleavage patterns from genomes.

Parallelly, the study on Rivulariapeptolide 1155 (Yang et al. 2019) introduced an innovative model for property prediction in drug discovery. Central to this model is a graph encoder architecture, previously applied in diverse domains like social networks and chemistry. This architecture's novelty lies in its hybrid representation, merging convolutions with descriptors to tailor task-specific encoding while embedding fixed descriptors as strong priors.

A notable feature of the model is its focus on bond-centered convolutions, which streamlines the message-passing algorithm by circumventing redundant loops. The model's efficacy is bolstered through Bayesian optimization for hyperparameter tuning and ensembling techniques to enhance accuracy, especially beneficial for complex data sets such as those in quantum mechanics.

Evaluations across various data sets have consistently demonstrated the model's superior performance, both in its default configuration and when optimized, underscoring its utility for chemists in drug discovery endeavors.

The discovery of Deepginsen (Merwin et al. 2020) via the DeepRiPP (Merwin et al. 2020) workflow is a testament to the synergy between machine learning and genomic analysis in natural product discovery. Deepginsen, a novel ribosomally synthesized natural product, was identified through a process that integrates multiple machine-learning technologies to bridge biosynthetic loci with their corresponding molecules.

The DeepRiPP workflow (Merwin et al. 2020), coupled with the NLPPrecursor deep learning framework, successfully navigated the challenge of distant precursor ORFs and tailoring enzymes. This framework excelled in identifying precursor peptides, paving the way for the correlation of specific ions to the gene cluster encoding Deepginsen. The subsequent in silico prediction and isolation efforts elucidated Deepginsen's structure, showcasing DeepRiPP's efficacy in streamlining the gene-to-molecule strategy for discovering novel RiPPs.

Deepginsen's design underscores the innovative capacity of DeepRiPP to automate the discovery of hidden natural products, merging advanced machine learning with genomic and metabolomics insights to explore the vast chemical diversity harbored within microbial genomes.

References

Amari, S.-I. (1993). Backpropagation and stochastic gradient descent method. *Neurocomputing* 5 (4–5): 185–196.

Born, M. and Oppenheimer, R. (1927). Zur quantentheorie der molekeln. *Annalen der Physik* 389 (20): 457–484.

Brockherde, F., Vogt, L., Li, L. et al. (2017). Bypassing the Kohn-Sham equations with machine learning. *Nature Communications* 8 (1): 1–10.

Chakraborti, N. (2004). Genetic algorithms in materials design and processing. *International Materials Reviews* 49 (3–4): 246–260.

Chmiela, S., Tkatchenko, A., Sauceda, H.E. et al. (2017). Machine learning of accurate energy-conserving molecular force fields. *Science Advances* 3 (5): e1603015.

David, L., Thakkar, A., Mercado, R., and Engkvist, O. (2020). Molecular representations in AI-driven drug discovery: a review and practical guide. *Journal of Cheminformatics* 12 (1): 1–22.

Delashmit, W.H., and Manry, M.T. (2005). *Recent Developments in Multilayer Perceptron Neural Networks*. Paper presented at the Proceedings of the seventh annual memphis area engineering and science conference, MAESC.

Deringer, V.L. and Csányi, G. (2017). Machine learning based interatomic potential for amorphous carbon. *Physical Review B* 95 (9): 094203.

Dietz, A. (1995). Yet another representation of molecular structure. *Journal of Chemical Information and Computer Sciences* 35 (5): 787–802.

Fang, X., Liu, L., Lei, J. et al. (2022). Geometry-enhanced molecular representation learning for property prediction. *Nature Machine Intelligence* 4 (2): 127–134.

Gaulton, A., Bellis, L.J., Bento, A.P. et al. (2012). ChEMBL: a large-scale bioactivity database for drug discovery. *Nucleic Acids Research* 40 (D1): D1100–D1107.

Gilmer, J., Schoenholz, S.S., Riley, P.F. et al. (2020). Message passing neural networks. In: *Machine Learning Meets Quantum Physics* (ed. K.T. Schütt, S. Chmiela, O.A. von Lilienfeld, et al.), 199–214. Springer.

Gu, J., Wang, Z., Kuen, J. et al. (2018). Recent advances in convolutional neural networks. *Pattern Recognition* 77: 354–377.

Hansen, K., Biegler, F., Ramakrishnan, R. et al. (2015). Machine learning predictions of molecular properties: accurate many-body potentials and nonlocality in chemical space. *Journal of Physical Chemistry Letters* 6 (12): 2326–2331.

Hansen, K., Montavon, G., Biegler, F. et al. (2013). Assessment and validation of machine learning methods for predicting molecular atomization energies. *Journal of Chemical Theory and Computation* 9 (8): 3404–3419.

Häse, F., Kreisbeck, C., and Aspuru-Guzik, A. (2017). Machine learning for quantum dynamics: deep learning of excitation energy transfer properties. *Chemical Science* 8 (12): 8419–8426.

He, K., Zhang, X., Ren, S., and Sun, J. (2016). *Deep residual learning for image recognition*. Paper presented at the Proceedings of the IEEE conference on computer vision and pattern recognition.

Heller, S., McNaught, A., Stein, S. et al. (2013). InChI-the worldwide chemical structure identifier standard. *Journal of Cheminformatics* 5 (1): 1–9.

Hermann, J., Schätzle, Z., and Noé, F. (2020). Deep-neural-network solution of the electronic Schrödinger equation. *Nature Chemistry* 12 (10): 891–897.

Hückel, E. (1931). Quantentheoretische beiträge zum benzolproblem. *Zeitschrift für Physik* 70 (3–4): 204–286.

Inc, D. (2020). Daylight Theory: SMARTS-A Language for describing molecular patterns. 2018. In.

Kearnes, S., McCloskey, K., Berndl, M. et al. (2016). Molecular graph convolutions: moving beyond fingerprints. *Journal of Computer-Aided Molecular Design* 30 (8): 595–608.

Kipf, T. N. and Welling, M. (2016). *Semi-Supervised Classification with graph convolutional networks*. arXiv preprint arXiv:1609.02907.

Kloosterman, A.M., Cimermancic, P., Elsayed, S.S. et al. (2020). Expansion of RiPP biosynthetic space through integration of pan-genomics and machine learning uncovers a novel class of lanthipeptides. *PLoS Biology* 18 (12): e3001026.

Krizhevsky, A., Sutskever, I., and Hinton, G.E. (2012). Imagenet classification with deep convolutional neural networks. In: *Advances in Neural Information Processing Systems*, vol. 25. Curran Associates, Inc.

Leigh, G.J. (2011). *Principles of Chemical Nomenclature: A Guide to IUPAC Recommendations*. Royal Society of Chemistry.

Lin, H., Ye, S., and Zhu, X. (2022). Geometry Orbital of Deep Learning (GOODLE): a uniform carbon potential. *Carbon* 186: 313–319.

Lin, H. and Zhu, X. (2023). Atomic fragment approximation from a tensor network. *Digital Discovery* 2 (6): 1688–1696.

Liu, T., Lin, Y., Wen, X. et al. (2007). BindingDB: a web-accessible database of experimentally determined protein–ligand binding affinities. *Nucleic Acids Research* 35 (Suppl. 1): D198–D201.

Mascarenhas, S., and Agarwal, M. (2021). *A comparison between VGG16, VGG19 and ResNet50 architecture frameworks for Image Classification*. Paper presented at the 2021 International conference on disruptive technologies for multi-disciplinary research and applications (CENTCON).

Merwin, N.J., Mousa, W.K., Dejong, C.A. et al. (2020). DeepRiPP integrates multiomics data to automate discovery of novel ribosomally synthesized natural products. *Proceedings of the National Academy of Sciences* 117 (1): 371–380.

Mullowney, M.W., Duncan, K.R., Elsayed, S.S. et al. (2023). Artificial intelligence for natural product drug discovery. *Nature Reviews Drug Discovery* 22 (11): 895–916.

Musil, F., Grisafi, A., Bartók, A.P. et al. (2021). Physics-inspired structural representations for molecules and materials. *Chemical Reviews* 121 (16): 9759–9815.

Pan, X., Wang, H., Li, C. et al. (2021). MolGpka: a web server for small molecule pK_a prediction using a graph-convolutional neural network. *Journal of Chemical Information and Modeling* 61 (7): 3159–3165.

Patra, T.K., Meenakshisundaram, V., Hung, J.-H., and Simmons, D.S. (2017). Neural-network-biased genetic algorithms for materials design: evolutionary algorithms that learn. *ACS Combinatorial Science* 19 (2): 96–107.

Pfau, D., Spencer, J.S., Matthews, A.G., and Foulkes, W.M.C. (2020). Ab initio solution of the many-electron Schrödinger equation with deep neural networks. *Physical Review Research* 2 (3): 033429.

Popescu, M.-C., Balas, V.E., Perescu-Popescu, L., and Mastorakis, N. (2009). Multilayer perceptron and neural networks. *WSEAS Transactions on Circuits and Systems* 8 (7): 579–588.

Qassim, H., Verma, A., and Feinzimer, D. (2018). *Compressed residual-VGG16 CNN model for big data places image recognition*. Paper presented at the 2018 IEEE 8th annual computing and communication workshop and conference (CCWC).

Ramakrishnan, R., Dral, P.O., Rupp, M., and Von Lilienfeld, O.A. (2014). Quantum chemistry structures and properties of 134 kilo molecules. *Scientific Data* 1 (1): 1–7.

Rowe, P., Deringer, V.L., Gasparotto, P. et al. (2020). An accurate and transferable machine learning potential for carbon. *The Journal of Chemical Physics* 153 (3): 034702.

Ruddigkeit, L., Van Deursen, R., Blum, L.C., and Reymond, J.-L. (2012). Enumeration of 166 billion organic small molecules in the chemical universe database GDB-17. *Journal of Chemical Information and Modeling* 52 (11): 2864–2875.

Schütt, K., Gastegger, M., Tkatchenko, A. et al. (2019). Unifying machine learning and quantum chemistry with a deep neural network for molecular wavefunctions. *Nature Communications* 10 (1): 1–10.

Schütt, K.T., Arbabzadah, F., Chmiela, S. et al. (2017). Quantum-chemical insights from deep tensor neural networks. *Nature Communications* 8 (1): 1–8.

Schütt, K.T., Glawe, H., Brockherde, F. et al. (2014). How to represent crystal structures for machine learning: towards fast prediction of electronic properties. *Physical Review B* 89 (20): 205118.

Schütt, K.T., Sauceda, H.E., Kindermans, P.-J. et al. (2018). SchNet—a deep learning architecture for molecules and materials. *The Journal of Chemical Physics* 148 (24): 241722.

Shelley, J.C., Cholleti, A., Frye, L.L. et al. (2007). Epik: a software program for pK a prediction and protonation state generation for drug-like molecules. *Journal of Computer-Aided Molecular Design* 21: 681–691.

Simonyan, K. and Zisserman, A. (2014). *Very deep convolutional networks for large-scale image recognition.* arXiv preprint arXiv:1409.1556.

Southan, C. (2013). InChI in the wild: an assessment of InChIKey searching in Google. *Journal of Cheminformatics* 5: 1–10.

Tsubaki, M. and Mizoguchi, T. (2020). Quantum deep field: data-driven wave function, electron density generation, and atomization energy prediction and extrapolation with machine learning. *Physical Review Letters* 125 (20): 206401.

Velickovic, P., Cucurull, G., Casanova, A., et al. (2017). Graph attention networks. *stat,* 1050(20), 10-48550.

Wang, R., Fang, X., Lu, Y. et al. (2005). The PDBbind database: methodologies and updates. *Journal of Medicinal Chemistry* 48 (12): 4111–4119.

Wang, Y., Xiao, J., Suzek, T.O. et al. (2009). PubChem: a public information system for analyzing bioactivities of small molecules. *Nucleic Acids Research* 37 (Suppl. 2): W623–W633.

Weininger, D. (1988). SMILES, a chemical language and information system. 1. Introduction to methodology and encoding rules. *Journal of Chemical Information and Computer Sciences* 28 (1): 31–36.

Weininger, D., Weininger, A., and Weininger, J.L. (1989). SMILES. 2. Algorithm for generation of unique SMILES notation. *Journal of Chemical Information and Computer Sciences* 29 (2): 97–101.

West, D.B. (2001). *Introduction to Graph Theory*, vol. 2. Upper Saddle River: Prentice Hall.

Wishart, D.S., Knox, C., Guo, A.C. et al. (2008). DrugBank: a knowledgebase for drugs, drug actions and drug targets. *Nucleic Acids Research* 36 (Suppl. 1): D901–D906.

Wiswesser, W.J. (1968). 107 years of line-formula notations (1861–1968). *Journal of Chemical Documentation* 8 (3): 146–150.

Wolpert, D.H. and Macready, W.G. (1997). No free lunch theorems for optimization. *IEEE transactions on evolutionary computation* 1 (1): 67–82.

Xiong, J., Li, Z., Wang, G. et al. (2022). Multi-instance learning of graph neural networks for aqueous pK_a prediction. *Bioinformatics* 38 (3): 792–798.

Xue, D., Balachandran, P.V., Hogden, J. et al. (2016). Accelerated search for materials with targeted properties by adaptive design. *Nature Communications* 7 (1): 1–9.

Yang, K., Swanson, K., Jin, W. et al. (2019). Analyzing learned molecular representations for property prediction. *Journal of Chemical Information and Modeling* 59 (8): 3370–3388.

Yang, X., Zhang, J., Yoshizoe, K. et al. (2017). ChemTS: an efficient python library for de novo molecular generation. *Science and Technology of Advanced Materials* 18 (1): 972–976.

Zhang, L., Han, J., Wang, H. et al. (2018). Deep potential molecular dynamics: a scalable model with the accuracy of quantum mechanics. *Physical Review Letters* 120 (14): 143001.

4

Autonomous Laboratory Empowered by AI and Robotics

4.1 Evolution of Laboratory

Chemists have spent substantial time on experimental tasks, such as synthesizing compounds, optimizing the reaction parameters, and identifying the components and molecular structure. These tedious works limit the creativity of chemists.

As the chemistry and material industry continues to evolve, the chemical industry has been working to discover new chemical reaction methods, catalysts, and more advanced equipment to reduce reaction times and improve laboratory efficiency. However, such evolution still requires human labor to get work in laboratories.

As introduced in previous chapters, rapid progress in artificial-intelligence (AI)-based algorithms and methods improves the efficiency of computational exploration of chemical space and helps design new materials. Yet, even the most advanced simulation technologies still cannot replace experimental data. Therefore, synthesis is the bottleneck in the materials design progress. We are demanding a complete rethinking of conventional approaches; applied AI and robotics methods in the laboratory is the path the researchers are pursing now.

Entering the twenty-first century, an increasing number of chemical literature has begun to mention the research on AI technology in chemical references. Previous researchers (Baum et al. 2021) utilized CAS content to study the growth and distribution of AI in related chemical publications from 2000 to 2020 (Figure 4.1). In 2012, AlexNet (Krizhevsky et al. 2012) introduced the CNN model into the Large-Scale Visual Recognition Challenge (LSVRC 2012) and achieved the first prize, pioneering deep neural network (DNN) models to solve practical problems. Scientists began to realize that with the development of computing power, deep learning methods could be used to attempt to solve various issues. Figure 4.1 shows an explosive growth in the number of chemical publication articles and patents related to AI in 2015.

4.2 Core Technologies in Autonomous Laboratories

4.2.1 Autonomous Laboratory Components

The widespread application and achievements of AI technology in the field of chemistry, including experimental aspects, underscore the transformative potential of AI

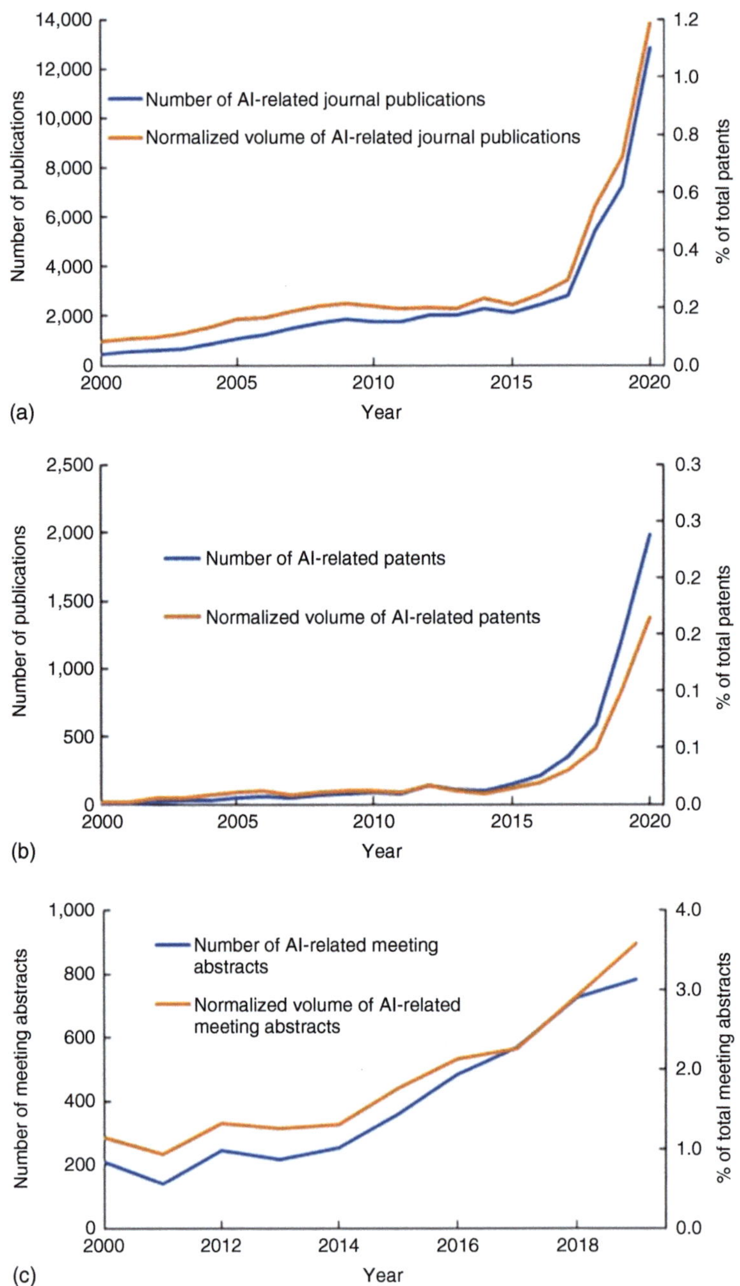

Figure 4.1 Annual publication volume in AI-related chemistry from 2000 to 2020: (a) Journal publications, (b) patent publications, and (c) ACS National Meeting abstracts. Source: Baum et al. (2021)/American Chemical Society.

in autonomous laboratories. Laboratories serve as physical spaces where AI, as a software algorithm, requires a medium to interact with the physical realm. Traditionally, AI planning followed by manual execution and manual data entry for AI analysis has been the norm. However, this approach is inefficient and faces developmental bottlenecks. Thus, the integration of robotics technology becomes essential. Robotics relies on digital signal control to execute various operations precisely and uses sensors to receive signals from the physical world for software analysis.

The significance of automation, even to the extent of autonomy in conducting experiments, has been recognized since Robert Bruce Merrifield, a Nobel Laureate in Chemistry in 1984 (Merrifield 1985), introduced his automated experimental apparatus during his Nobel lecture. This apparatus included a reaction vessel containing the resin with its growing peptide chain. Despite its innovation, this device was limited by the technological standards of its time and operated on manually set parameters – an automatic synthesis device. Over the subsequent four decades, collaboration among automation experts, computer scientists, chemists, and materials scientists has led to the development of various autonomous laboratory solutions.

Overall, an autonomous laboratory necessitates the fusion of AI technology with robotics to tailor design and customization according to the characteristics of chemistry and materials science laboratories. This integration forms a closed-loop logic essential for achieving autonomous experimentation. Such a closed-loop experimental process typically encompasses four main phases: design, react, test, and analyze.

The design phase represents a critical juncture within the entire workflow of autonomous laboratory systems, where the planning of experimental methodologies necessitates consideration of hardware capabilities, material constraints, and safety regulations. In various autonomous laboratory configurations, not all experiments may be feasible; thus, an ideal system should be able to automatically adjust to the hardware's capabilities, formulating experimental plans that can be safely and effectively executed. A distinguishing feature between autonomous and automatic laboratories lies in their operational approach: while automatic laboratories execute pre-defined experimental procedures, autonomous laboratories are capable of self-designing experiments. This includes optimizing and adjusting experimental parameters or even the experimental plans themselves based on prior results, thereby enabling self-improvement within the experimentation process. Consequently, the design phase demands adaptation based on the outcomes analyzed in previous phases.

The react phase executes the experiment's designed procedural phases through hardware. System designs vary according to specific reactions; some systems employ robotic arms with more than five degrees of freedom (DoF) to simulate human actions such as preparing reaction solutions, filtering, and transferring. Yoshikawa et al. introduced a motion planning scheme for a 7-DoF robotic arm (Yoshikawa et al. 2022), facilitating automation in chemistry laboratories. These systems typically offer greater flexibility for expansion by modifying robotic arm movements to accommodate additional reactors and equipment. However, modifications often require manual intervention, limiting the adaptability of the design phase. These robotic arms can also be mounted on mobile platforms, allowing operation across a broader range within laboratory spaces (Burger et al. 2020). Attempts have also been made to use mobile robotic arms in conjunction with state-of-the-art vision models to achieve more robust equipment utilization (Brohan et al. 2023); however, these models' stability and reliability remain uncertain.

An alternative strategy involves redesigning hardware to use a 3-axis robotic arm for moving reagent bottles and redesigning equipment used in various experimental steps like heating, mixing, and testing to be compatible with a 3-axis robotic arm. This hardware requires custom design but offers simplicity in action design and potential cost advantages in mass production compared to more complex robotic arms.

Another common hardware approach utilizes flow reactors where different liquid or gas reactants flow through a pipe with modules for heating, mixing, or testing incorporated along its length. This flow design is relatively fixed regarding reaction type; typically, one set of hardware can only conduct a limited range of experiments. Its advantage lies in the simplicity of design phases and reduced reactant volumes needed for exploratory experiments. However, scale-up effects may pose challenges as experiments viable in pipe-based setups may require adjustments or prove challenging to replicate in larger reaction vessels.

This exploration into autonomous laboratory components underscores the importance of integrating AI technology with robotics to innovate experimental methods across chemistry and materials science fields. By automating design and execution processes while adapting based on analytical feedback, these systems mark significant advancements toward achieving truly autonomous research laboratories.

In the testing phase, various analytical instruments are employed to assess the properties and states of post-reaction samples. The selection of measurement tools varies depending on the nature of the reaction under investigation, encompassing a range of techniques including Raman spectroscopy, infrared spectroscopy, chromatography, mass spectrometry, X-ray diffraction spectroscopy (XRD), nuclear magnetic resonance (NMR), and electrochemical workstations among others. Additionally, alternative instruments such as cameras with different spectral sensitivities might be utilized to observe sample colors under varying lighting conditions. Prior to characterization measurements, samples often undergo preparatory procedures like concentration, extraction, or freezing, which are integral components of the testing phase.

The analysis phase typically involves the application of software algorithms for collecting and interpreting data generated during the react and test phases. Within autonomous laboratories, an extensive volume of reaction process data can be gathered. For instance, simple temperature sensors within reaction vessels can monitor high-frequency temperature fluctuations. Such temperature trends can offer insights into reaction dynamics, revealing how variations in temperature profiles may influence reaction outcomes. This level of detailed information is challenging to capture in traditional laboratory settings and represents a broad area for exploration in autonomous laboratory systems. Data obtained from various characterization instruments during the testing phase require sophisticated algorithms for decoding essential information from raw data. A multitude of

AI techniques is applied to data analysis tasks. Traditional peak identification methods in spectroscopic analyses have utilized wavelet transforms and Fourier transforms. However, with advancements in AI over the past decade, deeper algorithms have been explored for recognition tasks across one-dimensional and two-dimensional spectra analyses, including convolutional neural networks (CNNs) for spectroscopic data and recurrent neural networks (RNNs) for chromatographic and temperature time-series data analysis. Furthermore, search algorithms and graph-based methods have been employed for interpreting more complex datasets such as those derived from mass spectrometry (e.g. CFM-ID technology (Allen et al. 2014; Djoumbou-Feunang et al. 2019; Wang et al. 2021)). These technological advancements enable researchers to move beyond labor-intensive analytical tasks toward more intelligent analytical processes.

Upon obtaining comprehensive analysis data, it becomes imperative to establish a closed-loop process that revisits the design phase. Initially described design phases involve crafting new experimental protocols based on prior knowledge. After completing the react–test–analyze phases, returning to the design step necessitates algorithms capable of generating subsequent experimental plans and parameters informed by previous data outcomes. Techniques such as reinforcement learning (RL) (Zhou et al. 2017), Bayesian algorithms (Shields et al. 2021), and genetic algorithms (Elliott et al. 2004) are instrumental in this context.

By integrating these methodologies within autonomous laboratories, researchers can achieve a seamless transition from experimental execution to insightful data analysis and back to refined experimental design. This cyclical process underscores the evolving nature of scientific inquiry facilitated by autonomous systems – a progression toward more efficient, accurate, and innovative materials chemistry research.

4.2.2 Reinforcement Learning

RL is a distinct category within the broader AI field that is particularly relevant to developing autonomous laboratories. At its core, RL is an area of machine learning that focuses on how agents can learn to make decisions by interacting with an environment to achieve a specific goal. Unlike supervised learning, where a model is trained with a labeled dataset, RL involves an agent that learns from the consequences of its actions through trial and error, receiving rewards or penalties based on the outcomes of its decisions.

One of the reasons RL is considered a crucial technology for autonomous laboratories is its ability to handle complex, dynamic environments. In a laboratory setting, numerous variables and uncertainties must be managed, such as changing experimental conditions, precise control of instruments, and the interpretation of diverse data types. RL algorithms can adapt to these changes and optimize experimental procedures by continuously learning from the environment.

RL employs a policy-based approach, where an agent learns a policy that maps states of the environment to the actions the agent should take. This policy is refined over time as the agent explores the environment and learns from the outcomes of its actions. This exploration–exploitation trade-off is a key aspect of RL, as the agent must balance trying new actions to discover better strategies (exploration) with leveraging its current knowledge to achieve the best results (exploitation).

In autonomous laboratories, RL can be used to automate various tasks, such as optimizing experimental protocols, controlling robotic systems for material synthesis, and even designing new materials with desired properties. For example, an RL agent can learn to adjust the parameters of a chemical reaction to maximize yield or purity based on feedback from analytical instruments. This speeds up the research process and reduces human error and the need for manual intervention.

RL is well suited for handling the sequential nature of experimental processes. Many laboratory experiments involve a series of steps that must be executed in a specific order, with each step depending on the outcomes of the previous ones. RL algorithms can learn to sequence these steps optimally, leading to more efficient and effective experimentation.

Another important aspect of RL in autonomous laboratories is its ability to deal with multi-objective optimization. In many research scenarios, there are multiple objectives to be balanced, such as maximizing the performance of a material while minimizing its cost or environmental impact. RL algorithms can learn to navigate these trade-offs and find solutions that best meet the overall research goals.

Furthermore, RL can facilitate the integration of different AI technologies in autonomous laboratories. For example, RL can work in conjunction with deep-learning models to process complex data, such as images or spectra, and make informed decisions based on that data. This integration enables the development of more sophisticated and intelligent autonomous systems.

The potential of RL in autonomous laboratories is not without challenges. One of the main hurdles is the need for large amounts of data and computational resources to train RL algorithms effectively. Additionally, ensuring the safety and reliability of RL-driven systems in a laboratory context is crucial, as errors could lead to hazardous situations or flawed research outcomes.

Despite these challenges, the ongoing advancements in RL and its integration with other AI technologies hold great promise for revolutionizing the way research is conducted in autonomous laboratories. By enabling more efficient, accurate, and innovative experimentation, RL is poised to play a key role in accelerating scientific discoveries and advancements in chemistry and materials science.

4.3 Example Autonomous Laboratory Solution

4.3.1 Automatic Device only Solution

Li et al. (2015) is a paper published in *Science* that presents an innovative automated synthesis platform capable of producing a wide range of small molecules, including complex natural product frameworks, using a uniform building-block-based approach and a novel catch-and-release chromatographic purification protocol. The platform strategically expands the scope of synthesis to include Csp3-rich polycyclic natural products by employing iterative Csp3 couplings and biosynthesis-inspired cascade polycyclizations. A key feature is the MIDA boronate-based purification method, which is applicable to all intermediates and enables streamlined automation and scalability.

A flow-based automated reaction device exemplifies the transition from manual experimentation to an automatic synthesis platform. This approach underscores the necessity of moving toward automation in synthesis, as demonstrated through specific case studies. Although the experimental schemes are not autonomously designed and lack automated data analysis and feedback optimization processes, they represent a significant step toward automation in chemical synthesis.

Similar initiatives, such as those reported by Masui et al. (2017), primarily focus on automating the reaction phase without extensively addressing the other three phases of autonomous laboratory operation. Rougeot et al. (2017) introduced a solution that achieves automated sampling of both soluble and slurry components in heterogeneous reactions, laying a hardware foundation for subsequent automated analysis. However, this solution does not extend to providing an automated analysis framework.

These examples highlight the evolving landscape of laboratory automation, where significant strides have been made in automating individual steps within the experimental process. Yet, there remains a critical need for comprehensive systems that encompass all aspects of autonomous operation, from experiment design and execution to data collection, analysis, and optimization based on feedback.

4.3.2 Solution that Including Design and React

Distinct from automatic synthesis platforms that merely serve as substitutes for manual experimentation, Chemputer system(Steiner et al. 2019) emerges as a compiler and robotic laboratory platform engineered for the synthesis of organic compounds through the utilization of standardized method descriptions. This innovative platform seamlessly integrates traditional laboratory apparatus, including round-bottom flasks, separatory funnels, and rotary evaporators, thereby ensuring compatibility with established literature protocols. Demonstrating its capability to conduct short syntheses of prevalent pharmaceuticals, the Chemputer system has achieved yields on par with those of manual synthesis processes. Noteworthy is its application in the synthesis of diphenhydramine hydrochloride, rufinamide, and sildenafil without the need for human intervention, marking a significant milestone in the transition from manual to automated chemical synthesis processes.

The system's Chempiler program generates specific instructions for the robotic platform, using a combination of GraphML for physical connectivity and ChASM (chemical assembly) language for machine operations. This setup allows the Chemputer to perform a wide range of syntheses by translating standardized synthesis procedures into executable instructions for the robotic platform, thereby automating the synthesis process from reaction setup to purification.

Moreover, the platform includes a feature for optimizing experimental procedures based on experimental data, although the article does not detail the specific algorithms used for this purpose. The ability to automatically adjust and optimize experimental protocols based on the outcomes of previous experiments represents a significant step toward fully autonomous chemical research laboratories. Chemputer systems are a major advance in the automation of chemical synthesis, fully defining a set of software to perform complex organic syntheses and designing the hardware to match.

RoboRXN from IBM, utilizing AI algorithms and a Flex-category automated-synthesis workstation, can automatically synthesize organic molecules by converting chemical recipes into machine-readable instructions. This system suggests retrosynthesis routes using AI to provide practical recipe options for producing specific molecules, emphasizing the potential to revolutionize chemical synthesis with AI and automation.

4.3.3 Solution in Reaction Optimization

Phoenics (Hase et al. 2018) is a method that combines Bayesian optimization with Bayesian kernel density estimation concepts. This algorithm is specifically designed for the optimization of experimental or computational procedures within chemistry, where objective evaluations are costly or time consuming. Phoenics is less sensitive to the optimization problem's response surface than other algorithms and showcases efficient parallel search capabilities. It balances exploration and exploitation strategies to identify optimal conditions for complex chemical reactions quickly. Figure 4.2 introduces the core concepts and operating mechanisms of Phoenics, including how to start from an unknown, possibly high-dimensional target function, process observed conditions through a Bayesian neural network (BNN), construct a surrogate model, and ultimately use the sampling parameter λ to adjust the strategy of exploring or exploiting the parameter space.

Phoenics stands out by proposing new experimental conditions based on all previous observations to avoid redundant evaluations and locate optimal conditions more efficiently. The algorithm employs BNNs to estimate kernel distributions associated with particular objective function values from observed parameter points, constructing a simple approximation to the objective function. This approach allows Phoenics to perform well across a broad range of chemical problems, efficiently optimizing unknown expensive-to-evaluate objective functions with minimal evaluations.

The paper also highlights Phoenics' ability to automatically adjust experimental plans based on previous data, effectively utilizing algorithms for automatic design, execution, data collection, and analysis of experiments. This functionality includes optimizing reaction conditions to achieve desired target dynamic behavior in chemical networks, showcasing Phoenics' utility in experimental chemistry applications where traditional optimization methods fall short due to the complexity and cost of evaluations. The limitation is that efficient parallelization strategies is not integrated with Phoenics.

Phoenics provides a powerful tool for optimizing experimental conditions and computational procedures. Its ability to navigate complex optimization landscapes with minimal evaluations positions it as a valuable asset for researchers seeking to

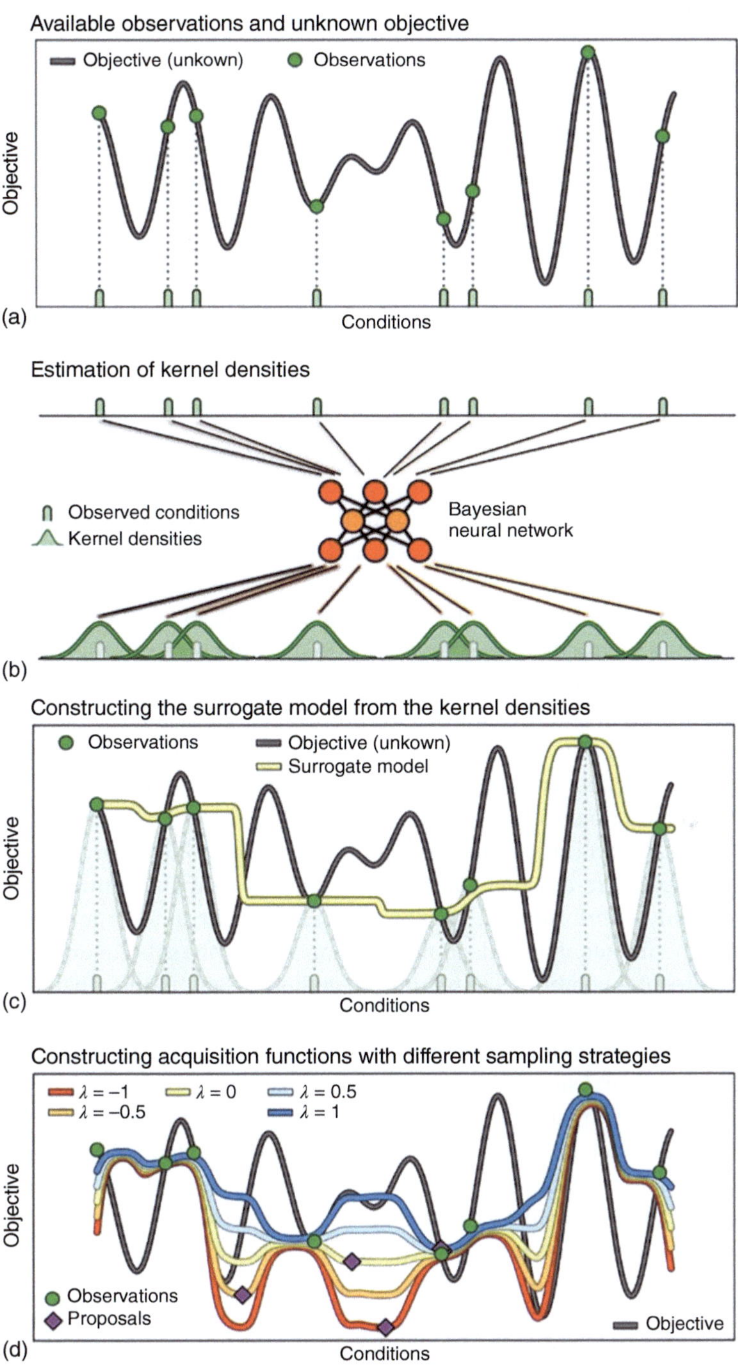

Figure 4.2 Illustration of the workflow of Phoenics. (a) Unknown, possibly high-dimensional, objective function of an experimental procedure or computation. (b) A Bayesian neural network processes the observed conditions, yielding a probabilistic model for estimating parameter kernel densities. (c) The surrogate model is constructed by weighting the estimated parameter kernel densities with their associated observed objective values. (d) The surrogate can be globally reshaped using a single sampling parameter λ to favor exploration (red) or exploitation (blue) of the parameter space. Source: Hase et al. (2018).

optimize chemical reactions and other complex systems within the constraints of limited resources.

4.3.4 Solutions Contain Full Phases

Since 2020, numerous scholars have proposed various autonomous laboratory solutions that fundamentally address the four phases previously mentioned: the design of reactions, execution, data collection and analysis, and the provision of closed-loop optimization strategies. Here are introductions to some exemplary solutions.

Burger et al. (2020) presents an innovative approach, leveraging a dexterous, free-roaming robot to autonomously conduct experiments aimed at improving photocatalysts for hydrogen production from water. The robot demonstrated an impressive capability to operate autonomously over eight days, carrying out 688 experiments within a ten-variable experimental space. It utilized a batched Bayesian search algorithm to navigate the experimental space, significantly enhancing the activity of photocatalyst mixtures compared to initial formulations. This approach increased the efficiency of experimental searches and demonstrated a modular strategy that could be applied in conventional laboratories for various research problems.

The introduced method setup featured a mobile robot capable of moving freely in a laboratory environment and dispensing insoluble solids and liquid solutions with high accuracy. This system operated using laser scanning and touch feedback for precise positioning, allowing it to perform tasks in complete darkness if required. The robot's ability to handle various materials and operate standard laboratory instruments without physical modifications showcases its versatility and adaptability to existing laboratory environments.

Integrating Bayesian optimization into the robotic workflow facilitated an efficient exploration of a complex, ten-dimensional chemical space, identifying optimal conditions for photocatalyst performance. This process involved the automatic design and execution of experiments, data collection, and analysis, and the subsequent optimization of experimental conditions based on the gathered data. Using a batched Bayesian search algorithm enabled the robot to make informed decisions on the next set of experiments to conduct, effectively balancing exploration and exploitation within the experimental space.

Nambiar et al. (2022) presents an innovative approach to optimizing a computer-aided synthesis planning (CASP)-proposed and human-refined multistep synthesis, using a modular robotic flow synthesis platform integrated with process analytical technology (PAT) for data-rich experimentation. The hardware design facilitates multi-step chemical synthesis including three reactions and one separation process, supported by multi-objective Bayesian optimization for the precise control of both categorical and continuous process variables. This setup enabled the automated execution of experiments, collection of analytical data, and optimization of reaction conditions, showcasing the seamless integration of automation, machine learning, and robotics in enhancing manual experimentation.

As illustration in Figure 4.3, this system includes all phases needed in autonomous laboratory: automatic experiment design, execution, data collection, and analysis.

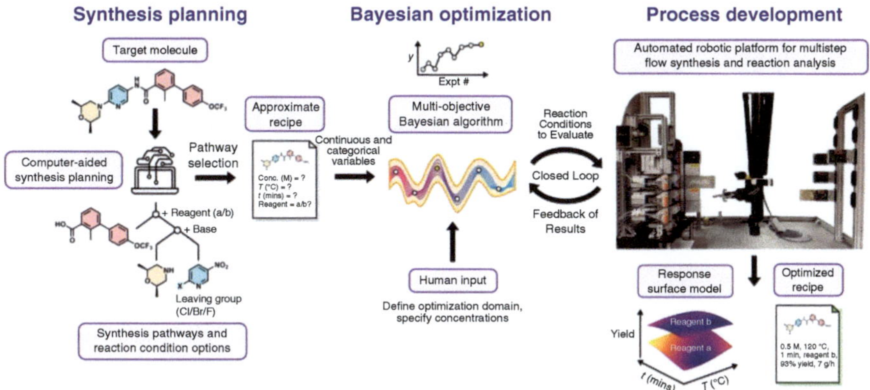

Figure 4.3 Overall approach for machine-assisted synthesis planning and process development. CASP recommendations for synthesis routes and reaction conditions to make humans assess the target molecule to create an approximate recipe with missing process details. A multi-objective Bayesian reaction optimization algorithm coupled to a robotic multistep flow synthesis platform optimizes continuous and categorical reaction conditions to fully specify the synthesis recipe. Source: Nambiar et al. (2022)/American Chemical Society.

The robotic platform's modularity and reconfigurability played a crucial role in optimizing residence times and order of addition for reactions, critical for minimizing undesired reactivity. The system employed liquid chromatography–mass spectrometry (LC-MS) device to collect data from the synthesized sample for further analysis. Using a multi-objective Bayesian optimization algorithm allowed for identifying optimal process conditions across the synthesis route, effectively addressing the challenges of automating multistep synthetic processes.

The automated optimization process included continuous and categorical variables optimization, leveraging an open-source Bayesian optimization package, Dragonfly, which employs a Gaussian process (GP) surrogate model. This model supported identifying optimal conditions for each step of the synthesis, including variables such as temperature, residence time, and stoichiometry, as well as categorical choices like solvent and catalyst type. The optimization demonstrated effective control over process conditions to maximize yield and productivity while minimizing cost.

Seifrid et al. (2022) development of an autonomous laboratory for synthesizing and analyzing new materials, focused on organic semiconductor lasers (OSLs). The lab integrates automated synthesis via iterative Suzuki–Miyaura cross-coupling reactions with analysis and purification capabilities, followed by optical characterization of molecules. It emphasizes the use of machine-learning (ML) and optimization algorithms to manage and analyze multidimensional data sets, aiming to accelerate molecular and materials discovery. The system is designed to execute experiments autonomously, including experiment planning, execution, data collection, and analysis. A notable limitation is the current focus on optical measurements in solution, acknowledging that material properties in the solid state are ultimately crucial.

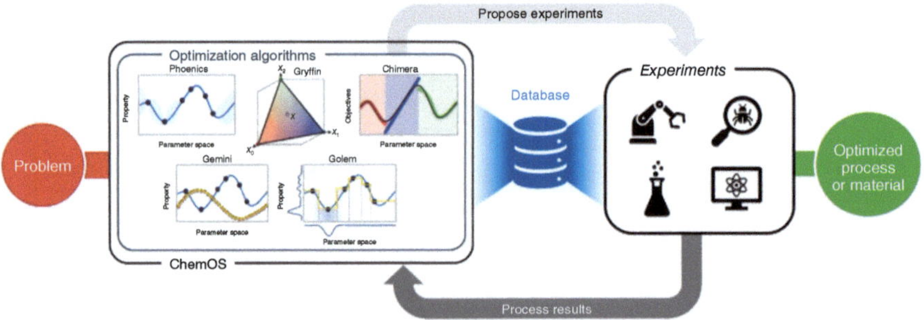

Figure 4.4 Material optimization workflow integrated with ChemOS and experiment devices. Source: From Seifrid et al. (2022).

The laboratory setup incorporates phases for experiment design, execution, data collection, and analysis. As shown in Figure 4.4, it utilizes ChemOS(Roch et al. 2020) for orchestration, ensuring efficient experiment scheduling and selection based on previous outcomes. The lab's capabilities extend to automated synthesis, purification, and optical characterization, aiming to streamline the discovery process for high-performing OSL materials. Challenges mentioned include the need for better software control and integration, handling heterogeneous systems, and adapting manual protocols for automated execution. Despite these obstacles, the self-driving lab represents a significant stride toward faster and more efficient materials discovery.

The system includes features for auto-optimizing experimental conditions based on collected data, leveraging algorithms like Bayesian optimization to refine experiment parameters continually. This approach enhances the lab's ability to discover and optimize materials efficiently, although the paper acknowledges challenges in achieving full autonomy, especially in integrating cognitive and motor functions seamlessly into the experimental workflow.

Hyper-converged Autonomous Organic Reaction Infrastructure (HAORI) (Xu et al. 2022) is another solution that contains all phases to archive autonomous laboratory, driven by the Spectrum Spiking Neural Network (SpecSNN) architecture as shown in Figure 4.5. HAORI is an integrated system encompassing reactions, characterization, and closed-loop optimization, aimed at synthesizing low dielectric constant polymers. It utilizes in situ spectroscopic feedback from automatic synthesis units to adjust reaction conditions dynamically, addressing the temporal discrepancies that often hinder optimization in traditional autonomous labs. Unlike previous systems, HAORI can improve both efficiency and accuracy by adapting reaction conditions in real-time based on spectroscopic data, making significant strides in autonomous laboratory capabilities.

HAORI comprises three main components: a potential functional material discovery algorithm, an automatic synthesis unit, and an in situ prediction and optimization system. The material discovery algorithm uses a random forest model for identifying materials with desired properties. The automatic synthesis unit, equipped with expert domain knowledge, performs parallel experiments

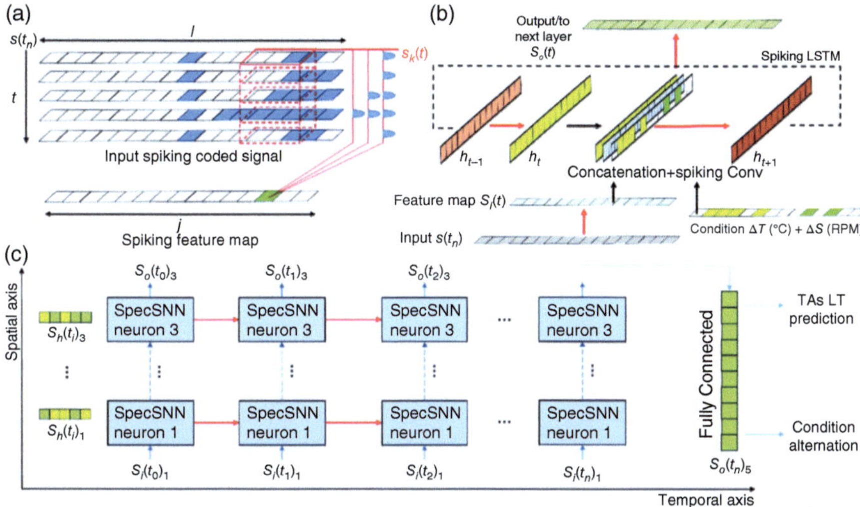

Figure 4.5 Structure of SpecSNN: the SNN-based experiment optimizer. (a) The progress of transferring spectrum data into pseudo-event-based rate coding feature maps. (b) A single SpecSNN cell's brief structure and functionality in the expanded view over time. (c) The structure of a 3-layer deep SpecSNN neural network. Source: Xu et al. (2022)/The Royal Society of Chemistry.

to optimize reaction conditions dynamically based on SpecSNN's output from real-time spectroscopic feedback. This system allows for the automatic design of experiments, execution, data collection, and analysis, enhancing the efficiency of synthesizing polymers with specific dielectric properties.

The implementation of SpecSNN enables HAORI to adjust experimental conditions automatically based on the analysis of spectroscopic data, optimizing the reaction pathway for better yields and properties. SpecSNN's utilization of spiking neural networks for real-time optimization represents a novel approach to handling the temporal aspects of chemical reactions, ensuring more accurate and efficient experimental outcomes. This optimization is critical in synthesizing materials with precise properties, such as low dielectric constants, which are essential for advancements in electronics and communications.

RoboChem (Slattery et al. 2024) is a benchtop robotic system designed to self-optimize, intensify, and scale-up of photocatalytic transformations. This system integrates liquid handlers, syringe pumps, a tunable continuous-flow photoreactor, inexpensive Internet of Things devices, and an in-line NMR system for real-time analysis. The core of RoboChem's operation lies in its closed-loop Bayesian optimization algorithm, which systematically explores a defined parameter space to identify optimal reaction conditions that maximize yield, throughput, or a combination of both. The platform operates autonomously, eliminating the need for extensive expertise in photocatalysis or scaling processes. It has demonstrated its ability across a diverse set of 19 molecules, covering various facets of photocatalysis like hydrogen atom transfer photocatalysis, photoredox catalysis, and metallaphotocatalysis. The Bayesian optimization algorithm effectively captures the intricate interdependencies among different reaction variables, consistently identifying optimal conditions that surpass manual approaches. The platform can scale up reactions within the same flow reactor system, with manual isolation processes applied to obtain meaningful quantities of pure compounds. The scalability is validated by the close alignment of isolated yields with NMR yields obtained by the platform, underlining its precision and reliability.

4.4 Advanced Autonomous Laboratory Solutions

As introduced in the preceding sections, numerous scholars have proposed their versions of autonomous laboratory systems. The year 2023 witnessed an explosive growth in generative AI, leading to the emergence of a plethora of innovative methods based on large models. Many of these methods hold potential for integration with autonomous laboratories. This section will introduce some of the latest technological perspectives on autonomous laboratories.

4.4.1 Advanced Experimental Data Analysis Methods

Vibrational spectroscopy techniques, such as Raman and Infrared (IR) spectroscopy, are pivotal in the characterization processes within autonomous laboratories. Raman spectroscopy, a photon-based technique, leverages the scattering of light to determine the energy shifts between incident and scattered photons, depicted on a spectrum with energy changes along the x-axis and intensity of scattered light on the y-axis. The location of a Raman band signifies the energy associated with molecular vibrations, with its intensity being directly proportional to the number of specific vibrations within the sample. Similarly, IR spectroscopy measures light absorption rather than scattering but operates under the same principle. Both Raman and IR spectroscopy are indispensable for molecule identification and characterization due to their sensitivity to molecular structure, offering unique vibrational spectra for each molecule. These methodologies find extensive application across diverse fields, including pharmaceuticals, materials science, and biology, for analyzing mixture compositions. Samples resulting from reactions are often complex mixtures; hence, dissecting these mixed spectra to identify various components is crucial for subsequent reaction analysis and optimization, providing valuable data for further experimentation.

The analysis of mixtures using Raman spectroscopy has recently been the subject of numerous studies aimed at developing reliable and efficient methods for accurately identifying mixture components. One approach that has been widely used is multivariate analysis, which involves the use of statistical algorithms such as Partial Least Squares (Zifarelli et al. 2020) (PLS), Principal Component Analysis (He et al. 2018) (PCA), or Soft Independent Modeling of Class Analogy (SIMCA) (Zhu et al. 2018), is a widely used approach for analyzing spectral data and identifying individual mixture components. However, this method has the disadvantage of requiring re-training of the model if new components are added to the mixture, which can be time consuming and resource intensive, and may limit the flexibility of the analysis. An alternative method is database query, in which the measured spectrum is compared to a database of known spectra to identify the present components. Standard metrics used to evaluate the similarity between the measured spectrum and spectra in the database include correlation, Euclidean distance (Taghizadeh et al. 2020), and Euclidean cosine (Samuel et al. 2021). The advantage of database query methods is that they do not require additional computational cost when new spectra are added to the database. The KnowItAll software, a well-known commercial spectrum analysis software, uses the database query method for component identification. However, this method is generally less effective for mixture analysis, as the computational cost can become prohibitively high for complex mixtures.

DNNs have gained significant attention in recent years due to their ability to achieve high performance in various applications, including computer vision (Chai et al. 2021) and natural language processing (Zhou et al. 2020). As a result, researchers have also explored using DNNs for analyzing and processing Raman spectra (Luo et al. 2022). Compared to traditional multivariate analysis methods, DNNs can often achieve higher prediction accuracy and better performance. However, it is essential to note that DNNs also have some limitations that need to be considered. One issue is that most DNN models need to be re-trained when applied to different datasets or when the model needs to be extended to new components since it is a multi-class classification model (Boateng et al. 2022). This process can be time consuming and may require much data to achieve good performance. One approach proposed to address this issue is using "component-specific" models, such as DeepCID(Fan et al. 2019), which trains a separate model for each component rather than a single model to predict the likelihood of all components. While this approach can reduce the need for re-training, the model must be trained for additional components when the analysis needs to be extended to new chemicals. In general, DNNs have been shown to achieve improved performance compared to traditional methods in many applications. However, it is essential to note that DNNs still have the disadvantage of requiring additional training when the model needs to be extended to new components, which can be time-consuming and may require a large amount of data to achieve good performance.

Tensor networks (TNs) are a potent tool for representing and analyzing multidimensional arrays, or tensors. They have garnered widespread use in fields such as quantum physics (Orús, 2019) and machine learning (Huggins et al. 2019; Li et al. 2021; Lin et al. 2021) due to their capability of compactly and efficiently capturing complex correlations and entanglement. Adopting the TN method for analyzing components in mixed spectra is a potential solution to improve the performance in spectrum analysis. Here, the Raman spectrum is picked as example to be illustrated, but it should be easily extended to other spectral data, even more measured data from devices in autonomous laboratory. TN method can identify potential components using only the wavenumber information of significant peaks and can be transferred to different databases with a single forward pass.

The molecular structure prediction module utilizes a transformer model for components do not present in the database to capture complex correlations, determine

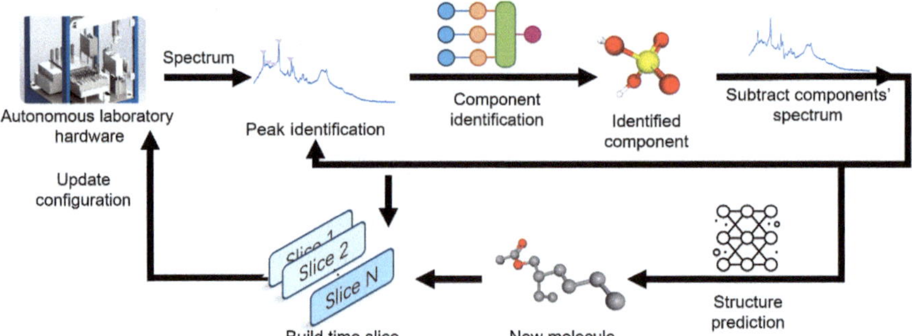

Figure 4.6 The measured spectrum undergoes processing to identify peaks, and all peak embeddings are passed to the TN model to query the components in the spectrum. The spectra of all known components are subtracted to obtain the remaining spectrum for continued identification. If the components cannot be identified through the TN, the molecular structure prediction module is applied to predict a possible new molecular structure.

molecular fragments, and identify possible fragment combinations to predict the detailed molecular structure.

In traditional database query method, the input spectrum is matched with all database spectra directly, and the possible component combinations are displayed in descending order of spectral similarity. However, using the complete spectrum for component matching is not a good option. Spectra are called the fingerprint of a molecule, meaning that for any molecular structure, its spectrum is unique. Moreover, we can only identify and match molecules by using a part of the spectrum. Additionally, the measurement of spectra is easily affected by various factors, such as baseline noise and intermolecular interactions, which can cause changes in the peak intensities of the spectra. In this case, using the entire spectrum for database matching will be interfered with by these additional factors, which will reduce the accuracy of the matching.

TN method does not use the complete spectrum as the input. Instead, as illustrated in Figure 4.6, it uses an algorithm to identify the N highest intensity peaks in the whole spectrum and takes their wavenumbers as inputs for the model. TN does not use the intensity of the peaks as input because, as mentioned earlier, various factors easily affect intensity. Based on this approach, the method could handle low signal-to-noise ratios, and we do not require high precision in baseline removal, nor do we need to use the intensity of the strong peaks as input.

The TN model identifies possible matching components and their confidence levels based on the input peaks. The spectrum of the matched components is subtracted from the original spectrum, and the identification process is repeated until the user-specified number of components is reached, or no significant peaks can be identified, or the confidence levels of all identified components are below the threshold. The remaining spectrum can be considered as an unknown component, and the molecular structure prediction module can be used to predict possible molecular structures.

Components identification. Various factors can influence the intensity of spectra peaks and are therefore not a reliable data source. Therefore, in our TN method, we only use wavenumbers to identify components. Generally, in the spectra of a mixture, at least a few peaks belong to the same component. In our process, after identifying one component, removing it makes it easier to identify the remaining components. Using only one peak for predicting components is clearly insufficient, as one peak can only roughly indicate one vibrational mode. Using two peaks with wavenumbers can also easily encounter situations where two component peaks in the database are close to each other. Therefore, at least three peaks need to be matched.

We employed a series of preprocessing methods to standardize the input spectra. To remove any offset or slope present in the spectrum, we used the asymmetric least squares (ALS) method for baseline removal, which has been proven effective. We then smoothed the spectra using a Savitzky–Golay filter to reduce noise and improve the signal-to-noise ratio. Additionally, we normalized the spectra using the min–max method to adjust their amplitudes to a standard scale. These preprocessing steps eliminate differences between spectrometers and remove any unwanted features that could interfere with the analysis. Significant peaks can be easily identified with a high-quality spectrum by comparing their intensity with their neighboring peaks.

The wavenumber of each peak is a scalar, which makes it susceptible to disturbance from external factors and solution effects. We use an embedding method to convert the scalar wavenumber to a high-dimensional space representation to make our method more robust. Small neural network (NN) model could be utilized to convert the scalar wavenumber into a d-dimensional vector. The NN model does not contribute to component identification; it enhances the performance of the following TN model by embedding the wavenumber into a high-dimensional space. The TN model is designed to capture correlations in high-dimensional space, and the embedded wavenumber provides a more robust representation.

Various types of TNs are commonly used for representing and manipulating high-dimensional arrays or tensors, including the canonical polyadic decomposition, Tucker TN, tensor train decomposition, and tensor ring (Evenbly and Vidal 2011). The Tucker TN has a central high-dimensional tensor composed directly from edge tensors; thus in our component identification application, the Tucker TN's central tensor is chosen for storing information from all components.

As an example, we use three most significant peaks, and for each pure component, the three most significant peaks are stored in the TN for identification. During the analysis of a mixture spectrum, more significant peaks (e.g. five peaks) need to be extracted. This results in C_3^5 combinations of peaks, which are all passed to the TN for prediction. All predicted component spectra are extracted from the database and compared with the mixture spectrum. The component that matches the most peaks with the mixture spectrum is determined as the identified component, and the spectrum of the identified component is subtracted from the mixture spectrum. The resulting spectrum has fewer vibration bands and is passed into the next iteration to identify other components. An example of the pipeline of extract component information from spectrum is shown in Figure 4.7.

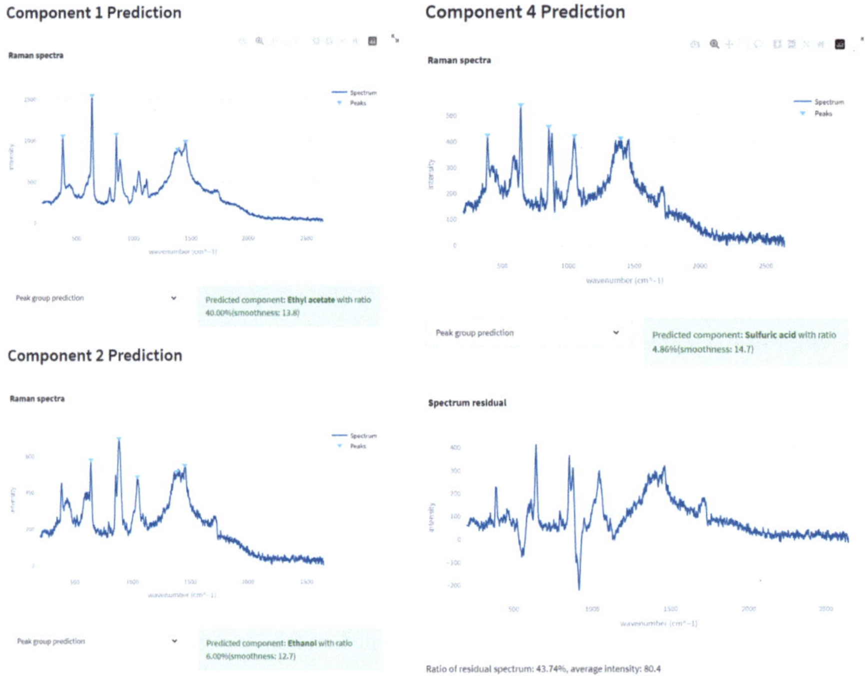

Figure 4.7 Visualization of step-by-step component identification by TN.

Previous ML methods are essentially multi-class models that require re-training for different datasets. Some DL models also require fine-tuning of the final classification layer, even if they do not necessitate re-training the entire model from scratch. The TN method we propose offers an advantage over traditional ML methods in that it does not require re-training the model for different databases. Like database query methods, the TN method requires only a single storage process and has the same benefits.

For predicting the residual spectra in the TN model or a new spectrum, the Molecular Structure Predictor is proposed to forecast potential molecular structures. The predictor estimates unknown molecules by anticipating the fragments present within the molecule and their arrangement. This is a complementary approach to the component identification model, enabling molecular analysis without reliance on a database.

The Molecular Structure Predictor, as shown in Figure 4.8, comprises two modules: The spectrum encoder, essentially models the correlation between fragments and spectra, facilitating the prediction of fragments existing within a molecule from its spectrum. The fragment decoder determines the final molecular structure by ascertaining the most probable connections between fragments through a granular analysis of the discrepancies between the input spectrum and the computed spectra of fragments. In this model, we leverage the entanglement properties of TNs and the attention mechanism of transformers to accomplish this fine-grained analysis.

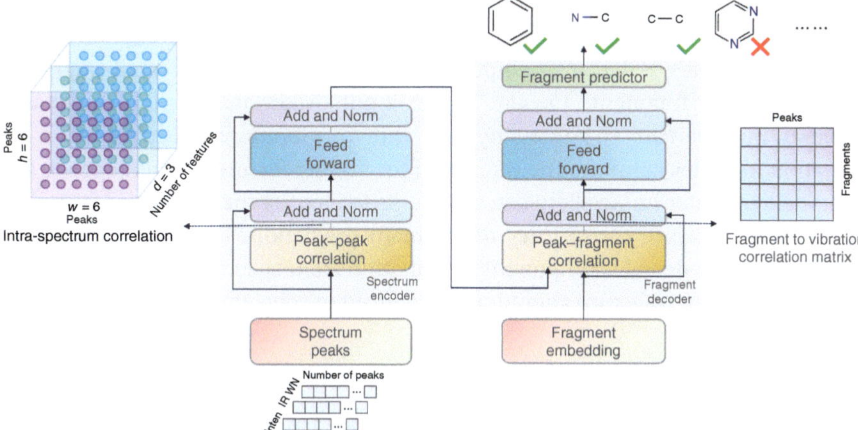

Figure 4.8 The molecular structure predictor encompasses two components: (a) Spectrum Encoder, which employs a spectral tensor network to capture the interrelation among peaks and project them onto a high-dimensional basis. This predictor calculates the entanglement between the input spectrum and fragment spectra to ascertain the constituent fragments within the input spectrum. (b) Fragment Decoder, which formulates all feasible molecular configurations from fragments, subsequently comparing the correlation between geometry and input spectrum to determine the target molecular structure.

In the spectrum encoder, the input vibrational spectra $s \in \mathbb{R}^{2 \times L}$, a matrix that includes 2-dimensional data (spectral intensity, peak wavenumber) for total L peaks in the spectra. The spectra data is projected to a d-dimensional space using a TN projector $T(\cdot) : \mathbb{R}^{2 \times L} \to \mathbb{R}^{d \times L}$. The high-dimensional representation for the spectrum would be passed into a sequence of peak-to-peak correlation modules $F(\cdot) : \mathbb{R}^{d \times L} \to \mathbb{R}^{d \times L}$, which updates the representation for each peak according to the inter-peak correlations. The same procedure is also applied to all fragments' spectra, to obtain fragments' feature $x_{\text{frag}} = F(S_{\text{frag}}) \in \mathbb{R}^{n \times d \times L}$, where n represents the number of fragments. The peak-to-peak correlation module F is designed to obtain the correlated weight from peak to peak. The high correlation weights indicate that the corresponding vibrations have overlapped atoms in the molecular structure.

The unknown and fragment spectra representation is fed to peak-to-fragment correlation module $G(\cdot) : (\mathbb{R}^{d \times L}, \mathbb{R}^{n \times d \times L}) \to \mathbb{R}^{d \times L}$, which obtain the correlation between unknown and fragments' features, to obtain the prediction for fragment existing in the unknown spectrum.

In the fragment decoder, the fragments predicted are combined into possible molecular graphs set $\mathcal{G} = \{g_1, \ldots, g_M\}$. All possible molecular graphs will be passed through the competitor, the graph with the highest logit value is the molecules with the closest correlation with the spectroscopy, i.e., the most likely molecular structure for the input spectrum.

Note that the molecular spectrum is not a mere summation of the individual fragment spectra. The internal correlation among fragments within a molecule accounts for this deviation. Analyzing such correlations proves challenging without density-functional theory (DFT) calculations, which is why identifying molecular structures solely from vibrational spectroscopy is complex. Employing the attention module, the proposed method utilizes fragment spectra as the foundation, examining the disparities between the input spectrum and the spectra of all fragments to determine the fragments present in the input. Additionally, other molecular fragments may perturb the intensity and shift of vibrational spectroscopy. The correlation modules compute the interrelations among features from all peaks and disentangle the interference between fragments to identify the molecular fragment. These modules ensure the efficacy of fragment prediction. Thus, why traditional method is hard to decode component information from spectrum, and TN method and transformer architecture should be introduced to solve them.

4.4.2 Design and Analysis in the Large Model Era

With the rapid development of large language models (LLMs), researchers have discovered the advantages of natural language interaction, notably its low barrier to entry. Individuals no longer need specialized training; they can simply communicate their experimental intentions to the system using natural language, potentially enabling specific functionalities. This advancement in large language models marks a significant milestone in the history of human–computer interaction. Consequently, we propose the integration of LLMs into autonomous laboratories.

Under the enhancement of LLMs, an ideal implementation of autonomous laboratories would allow users to convey their experimental desires through typed or spoken input. The system would then interpret this input using natural language processing technology and automatically convert it into scripts suitable for hardware execution.

Here is an example prompt from an LLM that illustrates how user experimental information described in natural language can be converted into formatted step-by-step instructions through integration with the model.

You are tasked with parsing experimental methods from research papers into clear and concise steps.

Your goal is to simplify the procedure into a series of straightforward steps, each denoted by a sequential number.

Avoid any hierarchical or sub-step formats like 1, a., b., 2, a., etc. Instead, use a straightforward numerical sequence: 1, 2, 3, and so on.

Focus on the essential actions and measurements in each step, such as amounts, time, temperature, atmosphere, and frequency.

Do not include additional information like the purpose of the step or background details.

Each step should contain a single operation.

If a step in the original method contains multiple operations, break it down into separate numbered steps.

Example:

Prepare a solution by dissolving x ml A in y ml B.

->

Add y ml B to a container.

Add x ml A to the same container.

Remember, your output should be a linear, numbered list of steps,

making the experiment easy to follow without any nested or complex hierarchical structures.

The steps are then formatted into scripts using the following prompts

Convert experimental methods into a graph format where each step is an edge and nodes represent locations (Reagent, Reactor, Temperature, Vibration, Atmosphere, Extraction).
Default to 'Reactor' when unsure. NO OTHER MODULE IS ALLOWED!
Index nodes sequentially as they appear.

Key Points:

The indices are by order of occurrence and are strictly matched.
The same reagents should come out from the same 'Reagent' index.
Keep track of the indices of the reagents used in the previous steps, make sure they match.
For conditions (Temperature, Vibration, Atmosphere) and Extraction, apply them directly to the Reactor if mentioned.
Stir is also Vibration.
Heat and cool are both Temperature.

Example:

Dissolve 1 ml FABr and 2 ml PbBr2 in 10 ml DMF separately as the precursor solutions.
Mix 1 ml octyl ammonium bromide and 2 ml oleic acid in 1-octadecene as the ligand solution.
Heat at 80 °C for 10 minutes and stir the precursor solution for 10 minutes.
Inject precursor solutions into the ligand solution.
->
[Reagent 1 -> Reactor 1] Add 10 ml DMF.
[Reagent 2 -> Reactor 1] Dissolve 1 ml FABr in 10 ml DMF.
[Reagent 1 -> Reactor 2] Add 10 ml DMF.
[Reagent 3 -> Reactor 2] Dissolve 2 PbBr2 in 10 ml DMF.

[Reagent 4 -> Reactor 3] Add 1 ml octyl ammonium bromide for mixing.
[Reagent 5 -> Reactor 3] Add 2 ml oleic acid for mixing.
[Temperature 1 -> Reactor 3] Heat the solution at 80 °C.
[Vibration 1 -> Reactor 3] Stir it for 10 minutes.
[Reactor 2 -> Reactor 3] Inject precursor solutions into the ligand solution.
[Reactor 3 -> Reactor 3] Inject precursor solutions into the ligand solution.

Similarly, in the analysis step, we aim to leverage large models for reading and analyzing characterization results, transforming them into natural language descriptions for the user. This approach is exemplified in the user interface depicted in Figure 4.12. Users can monitor the device via a digital twin model, supplemented by natural language descriptions and symbolic representations through this interface. This method enhances user comprehension and streamlines the communication of experimental statuses, progress, and analytical outcomes.

4.4.3 Experiment Visualization

Symbolic representation is strategically crafted based on the modular components of the experiment automation platform, as illustrated in Figure 4.9a. This figure outlines the workflow, starting with liquid reagents stored in bottles within the reagent storage area. These reagents are then extracted and injected into reactors using syringe pumps mounted on a 4-pipette hand. This hand moves atop a high-precision rail, allowing for adjustments in the distance between pumps via the liquid manipulation module. The chemical reactions occur within bottles placed in high-throughput reactors capable of heating, cooling, and oscillating at controllable parameters. Following the reaction phase, solutions can either be transferred using syringe pumps for further processing or subjected to postprocessing directly on the reactors. For postprocessing, modules such as those for photoluminescence (PL) and ultraviolet–visible (UV–vis) testing, as introduced in Figure 4.9a, can be attached to the reactors with a mechanical gripper. Available postprocessing options include both liquid–liquid/solid–liquid phase extraction and solid-phase extraction.

Figure 4.9b presents a symbolic representation of various experimental methods, demonstrating the system's modularity and flexibility. This approach allows for different experimental methods to be succinctly expressed through unique arrangements of symbols that correspond precisely to the automated experiment platform's modules.

Figures 4.10 and 4.11 showcase the foundational elements of this symbolic experiment visualization scheme and some actions combined from these elements, respectively. These figures highlight the scheme's adaptability and potential for refinement and expansion, making it applicable across diverse experimental setups and hardware configurations.

The visual representation scheme discussed here offers a more user-friendly alternative to the previously mentioned computer-oriented, formatted programming designs. It simplifies the user experience by avoiding lengthy textual descriptions and allows for easy customization of icons to represent different hardware configurations. This graphical user interface design approach represents

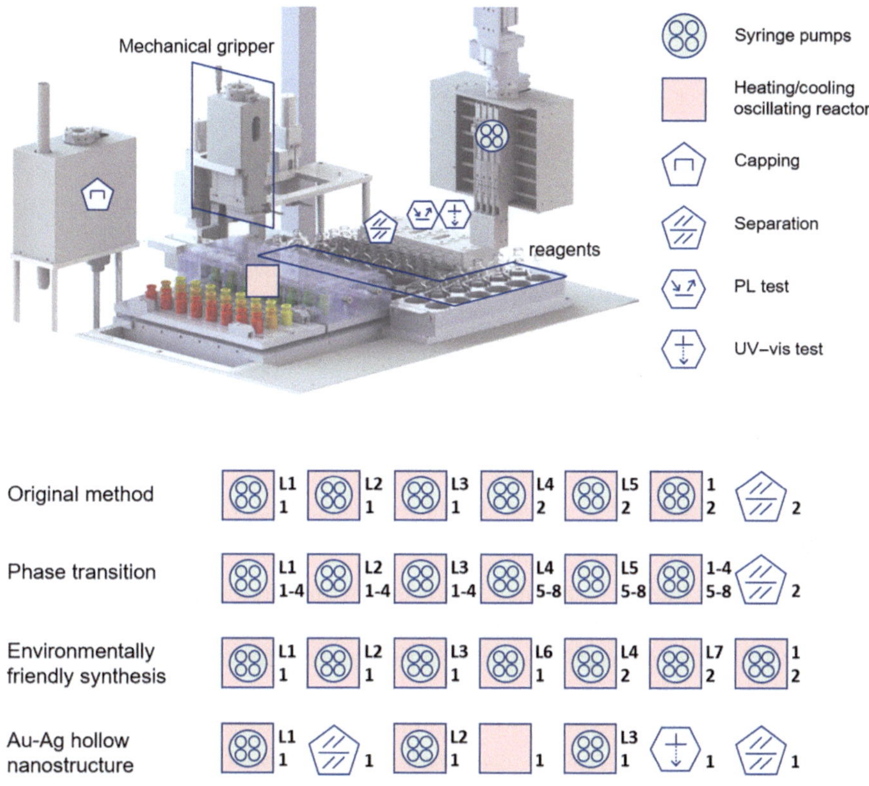

Figure 4.9 Symbolic representation of experiment schemes. In this representation, symbols are arranged chronologically, with each symbol depicted on a reactor-shaped background. Above the symbol, the required modules for that step are superimposed. If no reactor is involved, only the operation itself is depicted. Each step's reagents (or starting position) and target location are indicated on the right side. (a) Experiment automation platform modules. (b) Symbolic representations.

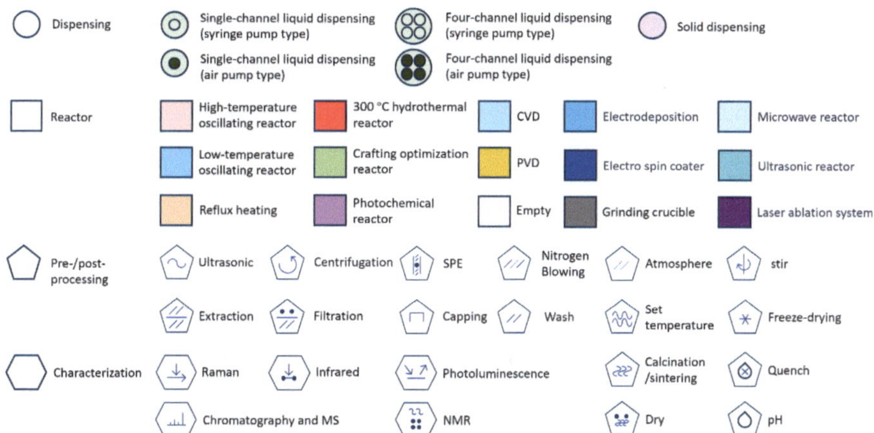

Figure 4.10 The base element for the symbol experiment visualization.

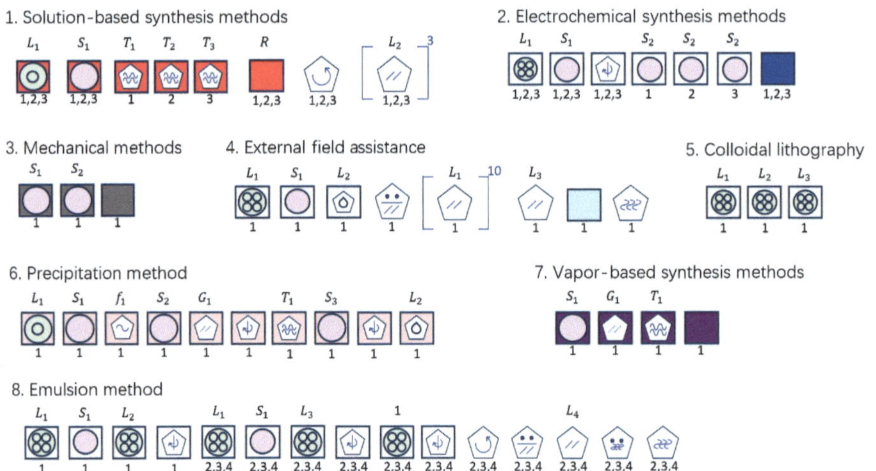

Figure 4.11 More examples of symbol experiment visualization, which shows the flexibility of this solution.

a future trend in human–computer interaction within autonomous laboratories. For well-developed autonomous laboratory solutions, users are not required to understand complex mechanisms or the intricacies of software and hardware, significantly lowering the barrier to entry. Therefore, an intuitive and straightforward user experience becomes crucial. An example of such a user interface is depicted in Figure 4.12, showcasing a symbolic representation of experimental schemes and extensively employing natural language to describe the current state of experiments, progress, and results of data analysis.

4.5 Future Prospects and Trends

The advent of autonomous laboratories empowered by AI and robotics marks a transformative era in scientific research, particularly in the realms of materials chemistry. This evolution signifies not merely an enhancement in the efficiency and precision of experimental procedures but heralds a paradigm shift toward a future where the boundaries of discovery are significantly expanded. The integration of AI and robotics into laboratory settings has initiated a journey from manual, labor-intensive processes to an era characterized by automation, innovation, and unprecedented scalability.

In envisioning the future development of autonomous laboratories, it is imperative to consider the trajectory of technological advancements alongside the evolving landscape of scientific inquiry. The convergence of AI, machine-learning algorithms, and robotics with sophisticated data analytics and cloud computing is poised to redefine the capabilities of research laboratories. These technologies enable the orchestration of complex experiments with minimal human intervention, allowing for the exploration of vast chemical spaces that were previously inaccessible due to practical limitations on time, resources, and human error.

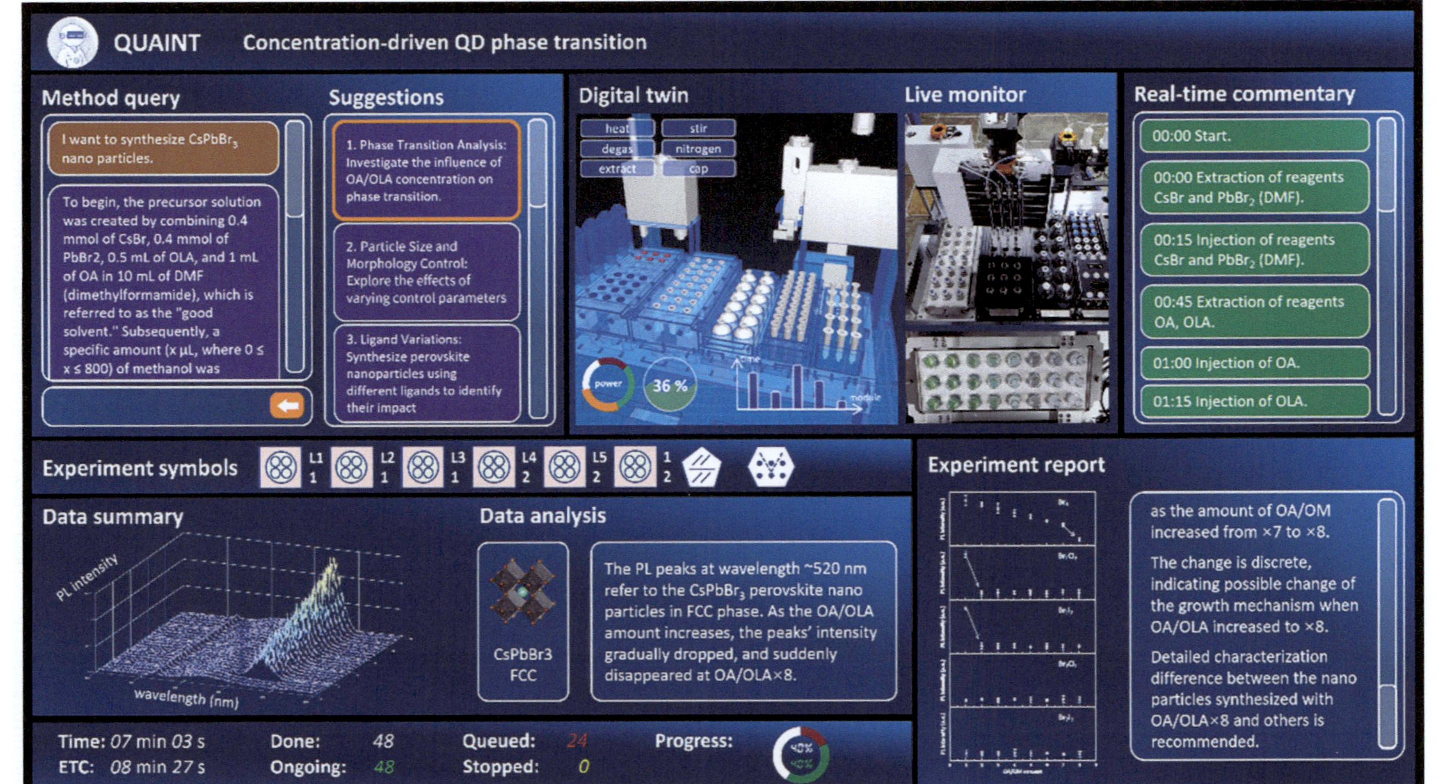

Figure 4.12 The user interface demo for the user to monitor the device through a digital twin model, nature language description, and symbolic representation.

One of the most significant long-term implications of this technological revolution is the democratization of scientific research. As autonomous laboratories become more prevalent and accessible, researchers around the globe will have unprecedented opportunities to engage in cutting-edge research without the need for extensive infrastructure or specialized training in experimental techniques. This democratization has the potential to accelerate scientific discoveries by enabling a broader base of researchers to contribute to solving complex global challenges such as energy sustainability, environmental protection, and healthcare innovation.

Moreover, the integration of AI and robotics into research laboratories promises to enhance the reproducibility and reliability of scientific findings. Automated systems can execute experiments with consistent precision and accuracy, reducing variability introduced by manual processes. This improvement in data quality is crucial for advancing our understanding of complex systems and phenomena, paving the way for more robust scientific theories and applications.

Another pivotal aspect of autonomous laboratories is their potential to foster interdisciplinary research. Automating experimental workflows liberates scientists from routine tasks, allowing them to focus on creative problem-solving across disciplinary boundaries. This collaborative approach is essential for addressing multifaceted problems that require converging perspectives from chemistry, physics, biology, engineering, and computational sciences.

As we look toward the horizon, it is clear that autonomous laboratories will play a central role in shaping the future landscape of materials chemistry research. These advanced systems will enable scientists to design and execute experiments with unprecedented complexity and precision. By harnessing the power of AI for predictive modeling and optimization, researchers will be able to discover new materials with tailored properties for applications ranging from renewable energy sources to advanced medical therapies.

In conclusion, autonomous laboratories represent a quantum leap forward in our quest for knowledge and innovation. The synergy between AI algorithms, robotics technology, and scientific inquiry holds immense promise for accelerating discoveries that can address some of society's most pressing challenges. As we continue to explore this exciting frontier, we as a global scientific community must ensure that these technologies are developed responsibly and equitably, fostering an inclusive environment where all researchers have access to the tools they need to contribute to our collective understanding of the natural world.

References

Allen, F., Pon, A., Wilson, M. et al. (2014). CFM-ID: a web server for annotation, spectrum prediction and metabolite identification from tandem mass spectra. *Nucleic Acids Research* 42 (W1): W94–W99.

Baum, Z.J., Yu, X., Ayala, P.Y. et al. (2021). Artificial intelligence in chemistry: current trends and future directions. *Journal of Chemical Information and Modeling* 61 (7): 3197–3212.

Boateng, D., Hu, C., Dai, Y. et al. (2022). Multicomponent Raman spectral regression using complete and incomplete models and convolutional neural networks. *Analyst* 147 (20): 4607–4615.

Brohan, A., Brown, N., Carbajal, J. et al. (2023). *Rt-2: Vision-Language-Action Models Transfer Web Knowledge to Robotic Control*. arXiv preprint arXiv:2307.15818.

Burger, B., Maffettone, P.M., Gusev, V.V. et al. (2020). A mobile robotic chemist. *Nature* 583 (7815): 237–241.

Chai, J., Zeng, H., Li, A., and Ngai, E.W. (2021). Deep learning in computer vision: a critical review of emerging techniques and application scenarios. *Machine Learning with Applications* 6: 100134.

Djoumbou-Feunang, Y., Pon, A., Karu, N. et al. (2019). CFM-ID 3.0: significantly improved ESI-MS/MS prediction and compound identification. *Metabolites* 9 (4): 72.

Elliott, L., Ingham, D., Kyne, A. et al. (2004). Genetic algorithms for optimisation of chemical kinetics reaction mechanisms. *Progress in Energy and Combustion Science* 30 (3): 297–328.

Evenbly, G. and Vidal, G. (2011). Tensor network states and geometry. *Journal of Statistical Physics* 145 (4): 891–918.

Fan, X., Ming, W., Zeng, H. et al. (2019). Deep learning-based component identification for the Raman spectra of mixtures. *Analyst* 144 (5): 1789–1798.

Hase, F., Roch, L.M., Kreisbeck, C., and Aspuru-Guzik, A. (2018). Phoenics: a Bayesian optimizer for chemistry. *ACS Central Science* 4 (9): 1134–1145.

He, X., Liu, Y., Huang, S. et al. (2018). Raman spectroscopy coupled with principal component analysis to quantitatively analyze four crystallographic phases of explosive CL-20. *RSC Advances* 8 (41): 23348–23352.

Huggins, W., Patil, P., Mitchell, B. et al. (2019). Towards quantum machine learning with tensor networks. *Quantum Science and Technology* 4 (2): 024001.

Krizhevsky, A., Sutskever, I., and Hinton, G.E. (2012). Imagenet classification with deep convolutional neural networks. *Advances in Neural Information Processing Systems* 25. https://proceedings.neurips.cc/paper/2012/hash/c399862d3b9d6b76c8436e924a68c45b-Abstract.html.

Li, J., Ballmer, S.G., Gillis, E.P. et al. (2015). Synthesis of many different types of organic small molecules using one automated process. *Science* 347 (6227): 1221–1226.

Li, J., Liu, R., Lin, H. et al. (2021). Tensor network-encrypted physical anti-counterfeiting passport for digital twin authentication. *ACS Applied Materials & Interfaces* 13 (51): 61536–61543.

Lin, H., Ye, S., and Zhu, X. (2021). *Tensor Network for Supervised Learning at Finite Temperature*. arXiv preprint arXiv:2104.05439.

LSVRC, I. (2012). Large Scale Visual Recognition Challenge 2012 (ILSVRC2012). Disponible on-line: https://www.image-net.org/challenges/LSVRC/2012/ (último acceso: Marzo 2021).

Luo, R., Popp, J., and Bocklitz, T. (2022). Deep learning for Raman spectroscopy: a review. *Analytica* 3 (3): 287–301.

Masui, H., Yosugi, S., Fuse, S., and Takahashi, T. (2017). Solution-phase automated synthesis of an α-amino aldehyde as a versatile intermediate. *Beilstein Journal of Organic Chemistry* 13 (1): 106–110.

Merrifield, R.B. (1985). Solid phase synthesis (Nobel lecture). *Angewandte Chemie International Edition in English* 24 (10): 799–810.

Nambiar, A.M., Breen, C.P., Hart, T. et al. (2022). Bayesian optimization of computer-proposed multistep synthetic routes on an automated robotic flow platform. *ACS Central Science* 8 (6): 825–836.

Orús, R. (2019). Tensor networks for complex quantum systems. *Nature Reviews Physics* 1 (9): 538–550.

Roch, L.M., Häse, F., Kreisbeck, C. et al. (2020). ChemOS: an orchestration software to democratize autonomous discovery. *PLoS One* 15 (4): e0229862.

Rougeot, C., Situ, H., Cao, B.H. et al. (2017). Automated reaction progress monitoring of heterogeneous reactions: crystallization-induced stereoselectivity in amine-catalyzed aldol reactions. *Reaction Chemistry & Engineering* 2 (2): 226–231.

Samuel, A.Z., Mukojima, R., Horii, S. et al. (2021). On selecting a suitable spectral matching method for automated analytical applications of Raman spectroscopy. *ACS Omega* 6 (3): 2060–2065.

Seifrid, M., Pollice, R., Aguilar-Granda, A. et al. (2022). Autonomous chemical experiments: challenges and perspectives on establishing a self-driving lab. *Accounts of Chemical Research* 55 (17): 2454–2466.

Shields, B.J., Stevens, J., Li, J. et al. (2021). Bayesian reaction optimization as a tool for chemical synthesis. *Nature* 590 (7844): 89–96.

Slattery, A., Wen, Z., Tenblad, P. et al. (2024). Automated self-optimization, intensification, and scale-up of photocatalysis in flow. *Science* 383 (6681): eadj1817.

Steiner, S., Wolf, J., Glatzel, S. et al. (2019). Organic synthesis in a modular robotic system driven by a chemical programming language. *Science* 363 (6423): eaav2211.

Taghizadeh, A., Leffers, U., Pedersen, T.G., and Thygesen, K.S. (2020). A library of ab initio Raman spectra for automated identification of 2D materials. *Nature Communications* 11 (1): 1–10.

Wang, F., Liigand, J., Tian, S. et al. (2021). CFM-ID 4.0: more accurate ESI-MS/MS spectral prediction and compound identification. *Analytical Chemistry* 93 (34): 11692–11700.

Xu, Y., Qiu, K., Xiao, S. et al. (2022). Hyperconverged autonomous organic reaction infrastructure (HAORI) driven by SpecSNN, for low dielectric constant polymer research. *Digital Discovery* 1 (4): 375–381.

Yoshikawa, N., Li, A.Z., Darvish, K. et al. (2022). *An Adaptive Robotics Framework for Chemistry Lab Automation*. arXiv preprint arXiv:2212.09672.

Zhou, M., Duan, N., Liu, S., and Shum, H.-Y. (2020). Progress in neural NLP: modeling, learning, and reasoning. *Engineering* 6 (3): 275–290.

Zhou, Z., Li, X., and Zare, R.N. (2017). Optimizing chemical reactions with deep reinforcement learning. *ACS Central Science* 3 (12): 1337–1344.

Zhu, L., Sun, J., Wu, G. et al. (2018). Identification of rice varieties and determination of their geographical origin in China using Raman spectroscopy. *Journal of Cereal Science* 82: 175–182.

Zifarelli, A., Giglio, M., Menduni, G. et al. (2020). Partial least-squares regression as a tool to retrieve gas concentrations in mixtures detected using quartz-enhanced photoacoustic spectroscopy. *Analytical Chemistry* 92 (16): 11035–11043.

5

Large Language Models for the Autonomous Material Research

5.1 Review of Large Language Models Development and Applications

This chapter aims to delve into the application of Natural Language Processing (NLP), with a particular focus on Large Language Models (LLMs) for Autonomous Material Research. It encompasses a comprehensive overview, structured around three core sections:

1) A concise review of the development of NLP and LLMs, highlighting the evolutionary trajectory and key milestones that have shaped the current landscape.
2) An examination of the application of LLMs in Material Science research over the past two years, reviewing significant advancements and the impact of these technologies in accelerating research and discovery in the field.
3) A presentation of our team's contributions and achievements in this area, showcasing how our work has leveraged LLMs to push the boundaries of what's possible in Material Science research.

Within the broad scope of artificial intelligence (AI) application across various fields, the role of NLP has become increasingly crucial, gaining widespread acceptance. The field of NLP has made significant progress, particularly with the advent of transformer technology, yet applying these advancements to specific scholarly tasks continues to present challenges. The utilization of pretrained models and publicly available datasets in a range of applications faces limitations; current resources cannot be directly and effectively applied to specialized tasks like analyzing the content of academic papers without considerable adjustments. The shortcomings of unsupervised NLP classification methods, such as Word2Vec (Goldberg and Levy 2014), become apparent in their failure to precisely capture the complex essence of scholarly content. These techniques often highlight superficial similarities at the expense of deeper, research-specific characteristics, including thematic focus, methodological approaches, and argumentative logic. The development of an AI assistant designed to synthesize and predict research outcomes requires an approach that aligns with the analytical capabilities of human researchers. This includes a focus on figures and their captions, recognized for their

AI and Robotic Technology in Materials and Chemistry Research, First Edition. Xi Zhu.
© 2025 WILEY-VCH GmbH. Published 2025 by WILEY-VCH GmbH.

rich informational value, alongside abstracts and conclusions, to capture the full breadth of scholarly communication.

Conventional NLP methods have historically concentrated on textual data, neglecting the insights that can be gleaned from figures. However, the introduction of Vision Transformer (ViT (Radford et al. 2021)) and subsequent innovations have paved the way for methods that incorporate multi-head self-attention mechanisms. These advancements enable the simultaneous analysis of text and visual data, offering a holistic view within a singular analytical framework. The emergence of multi-modal learning has led to the creation of models like Vision and Language BERT (ViLBERT (Lu et al. 2019)) and VisualBERT (Li et al. 2019). These models are adept at handling tasks that necessitate the merging of visual and textual data, such as visual question answering (VQA) and image captioning (IC). A notable challenge in adapting these models for academic research is the similarity in image features across documents within the same research area. Figures, whether they depict data or results, tend to share stylistic and content similarities within specific fields. Addressing this issue requires a comprehensive approach that considers both figure and text features in the creation and ongoing update of databases, ensuring a dynamic link between these two elements. Despite the advances in multi-modal methodologies, achieving a seamless integration of textual and visual elements in academic literature remains a formidable task, underscoring the necessity for continued enhancement and innovation in this domain.

Since the year 2022, the domain of scientific exploration has undergone a significant transformation due to the introduction and assimilation of LLMs like ChatGPT (Brown et al. 2020). This development heralds the beginning of a novel phase in scholarly research and methodologies, illuminating two crucial aspects of the evolution of research tools and their potential to shape the future of scientific discoveries. Figure 5.1 illustrates the capabilities of LLMs, concluded by Minaee et al. (2024). The advent of LLMs has brought about a profound impact on the academic world, revolutionizing the methods of research assistance and the digestion of knowledge. These models have exhibited exceptional capabilities in understanding and generating text that mimics human conversation, thereby becoming indispensable across a myriad of scientific research applications. Ranging from material science (Olivetti et al. 2020) to a variety of interdisciplinary fields, LLMs provide a powerful foundation for boosting research efficiency and fostering innovation. The widespread adoption of AI-assisted research tools following the launch of ChatGPT (Inciteful 2023; Ought 2023; PBC 2023), together with platforms like scite (Nicholson et al. 2021) integrating generative AI for enhancing query services (scite 2023), signifies a crucial juncture in the progression of research methodologies.

Next, some reported applications of LLM in material research will be analyzed. The first and most popular application is predictive modeling, which typically includes various property prediction (Dinh et al. 2022), reaction prediction, and further. Predictive modeling serves as a prevalent use case for machine learning within the field of chemistry, particularly for predicting material characteristics. Utilizing the Language-Interfaced Fine-Tuning (LIFT) approach (Dinh et al. 2022), Jablonka et al. (2023b) demonstrated the capability of LLMs to forecast a variety

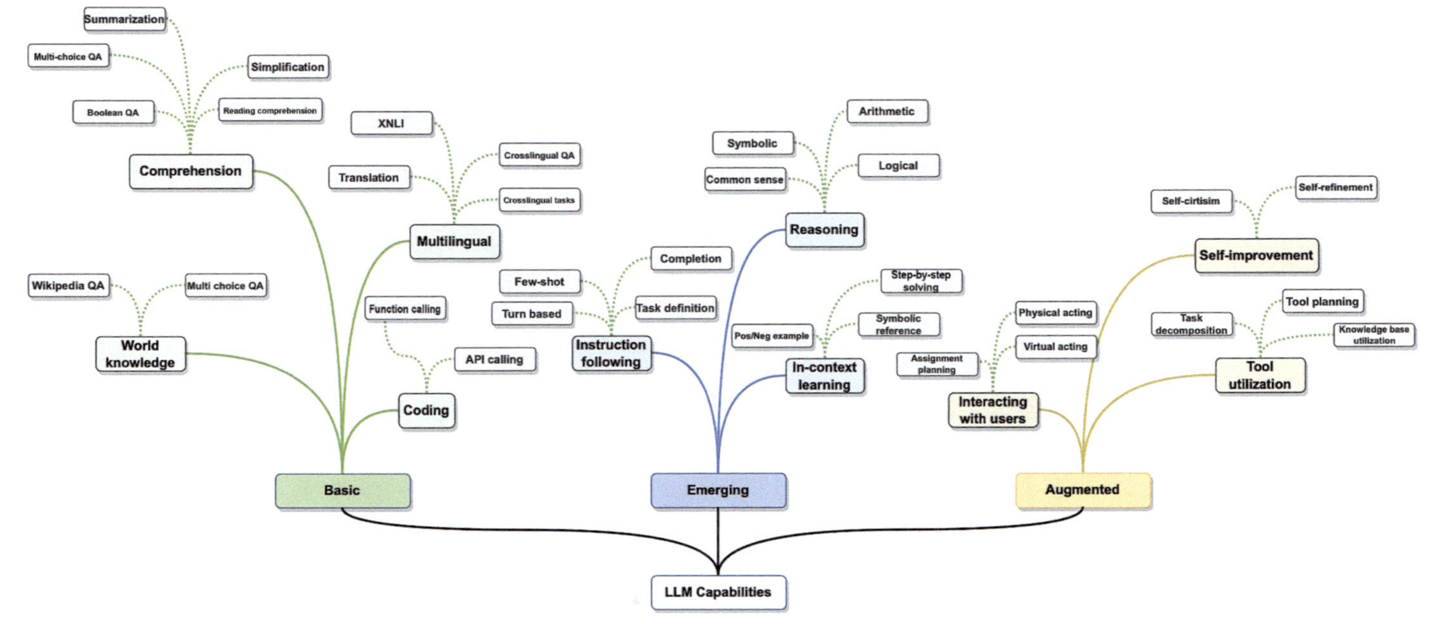

Figure 5.1 Capabilities of LLM. Source: Minaee et al. (2024)/arXiv/CC-BY-4.0.

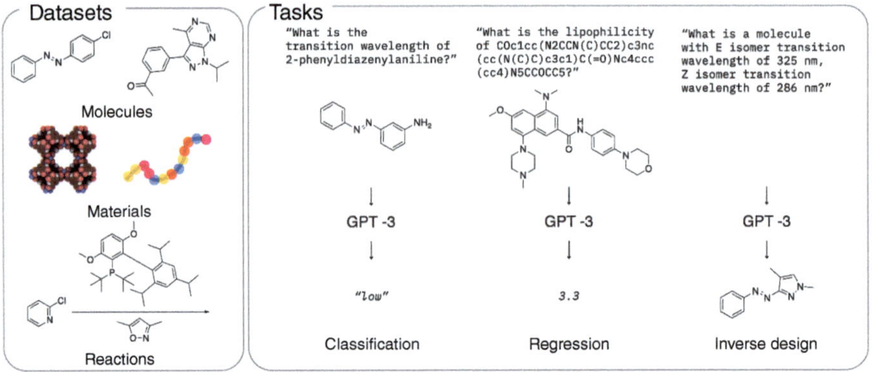

Figure 5.2 Overview of the datasets and tasks of LLMs predictive modeling. In their study, the authors evaluate GPT-3's performance across datasets that cover a broad spectrum of chemical entities, including molecules, materials, and reactions. They explore a variety of tasks within these datasets, such as classification, where they predict categories (for instance, "high" or "low") based on text descriptions of molecules, materials, or reactions. Additionally, they delve into regression tasks, which involve forecasting numerical values, and the inverse design, focusing on the generation of molecular structures. Source: Jablonka et al. (2023b)/ChemRxiv/CC-BY-4.0.

of chemical attributes, such as energy, solubility or the HOMO–LUMO gaps, employing molecular line representations like self-referencing embedded strings (SELFIES) (Krenn et al. 2022, 2020) and Simplified Molecular Input Line Entry System (SMILES). Selected tasks and results can be seen in Figure 5.2. Advancing this concept, Ramos et al. (2023) applied the same framework, incorporating in-context learning (ICL), to perform Bayesian optimization. This method guides experimental procedures without the necessity for model training, which can produce results in a straightforward way, as shown in Figure 5.3.

Furthermore, LLMs have shown potential in facilitating Molecule Discovery through Context. The contextual word embeddings generated by an LLM, tailored to the specific sequence or sentence in which they appear, can more adeptly understand the contextual nuances of words within a corpus (Selva Birunda and Kanniga Devi 2021). Drawing on this concept, researchers at GlobusLabs, including Zhi Hong and Logan Ward, explored whether similar embeddings could aid in identifying hydrogen carrier molecules, crucial for energy storage solutions. They utilized the ScholarBert model (Hong et al. 2022), which was trained on an extensive collection of scientific literature provided by the Public.Resource.Org nonprofit. For various potential molecules, the team identified sentences within the Public.Resource.Org dataset and employed the mean of these sentence embeddings as a molecular fingerprint. By comparing these fingerprints, they were able to prioritize molecules based on their resemblance to established hydrogen carrier molecules. A visual examination confirmed that the molecules selected shared characteristics with known hydrogen carriers.

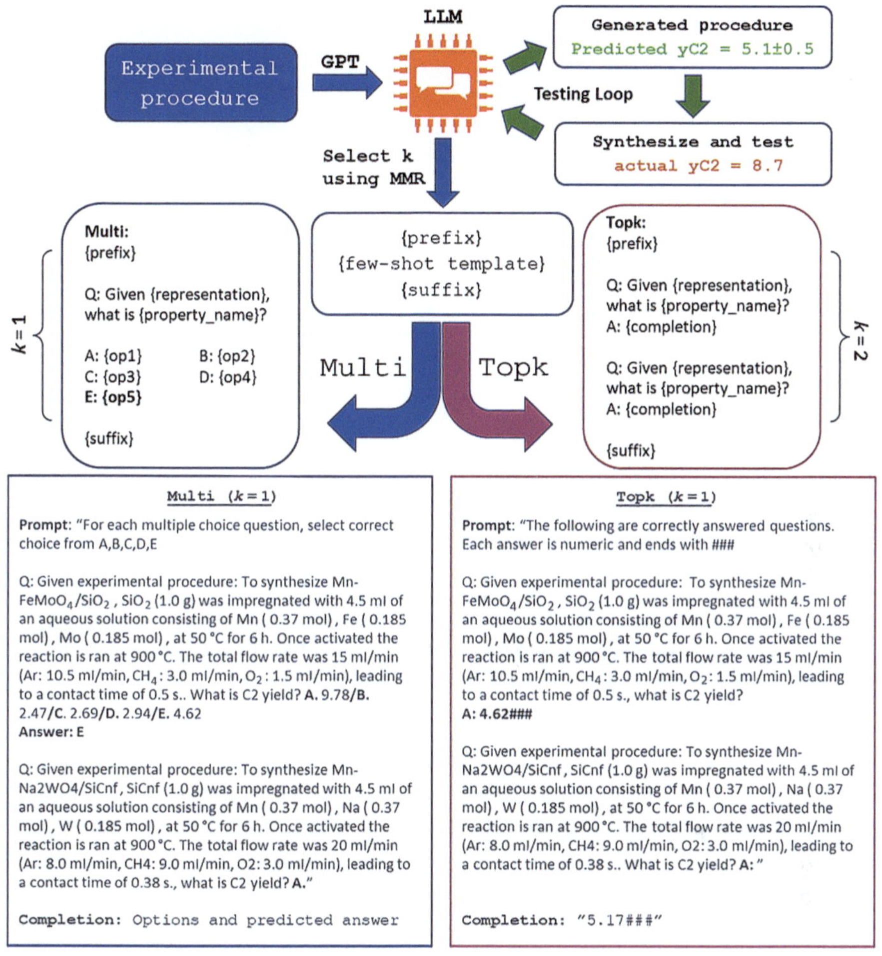

Multi $(k=1)$

Prompt: "For each multiple choice question, select correct choice from A,B,C,D,E

Q: Given experimental procedure: To synthesize Mn-FeMoO$_4$/SiO$_2$, SiO$_2$ (1.0 g) was impregnated with 4.5 ml of an aqueous solution consisting of Mn (0.37 mol), Fe (0.185 mol), Mo (0.185 mol), at 50 °C for 6 h. Once activated the reaction is ran at 900 °C. The total flow rate was 15 ml/min (Ar: 10.5 ml/min, CH$_4$: 3.0 ml/min, O$_2$: 1.5 ml/min), leading to a contact time of 0.5 s.. What is C2 yield? **A.** 9.78/**B.** 2.47/**C.** 2.69/**D.** 2.94/**E.** 4.62
Answer: E

Q: Given experimental procedure: To synthesize Mn-Na2WO4/SiCnf, SiCnf (1.0 g) was impregnated with 4.5 ml of an aqueous solution consisting of Mn (0.37 mol), Na (0.37 mol), W (0.185 mol), at 50 °C for 6 h. Once activated the reaction is ran at 900 °C. The total flow rate was 20 ml/min (Ar: 8.0 ml/min, CH4: 9.0 ml/min, O2: 3.0 ml/min), leading to a contact time of 0.38 s., what is C2 yield? **A.**"

Completion: Options and predicted answer

Topk $(k=1)$

Prompt: "The following are correctly answered questions. Each answer is numeric and ends with ###

Q: Given experimental procedure: To synthesize Mn-FeMoO$_4$/SiO$_2$, SiO$_2$ (1.0 g) was impregnated with 4.5 ml of an aqueous solution consisting of Mn (0.37 mol), Fe (0.185 mol), Mo (0.185 mol), at 50 °C for 6 h. Once activated the reaction is ran at 900 °C. The total flow rate was 15 ml/min (Ar: 10.5 ml/min, CH$_4$: 3.0 ml/min, O$_2$: 1.5 ml/min), leading to a contact time of 0.5 s., what is C2 yield?
A: 4.62###

Q: Given experimental procedure: To synthesize Mn-Na2WO4/SiCnf, SiCnf (1.0 g) was impregnated with 4.5 ml of an aqueous solution consisting of Mn (0.37 mol), Na (0.37 mol), W (0.185 mol), at 50 °C for 6 h. Once activated the reaction is ran at 900 °C. The total flow rate was 20 ml/min (Ar: 8.0 ml/min, CH4: 9.0 ml/min, O2: 3.0 ml/min), leading to a contact time of 0.38 s.. What is C2 yield? A:"

Completion: "5.17###"

Figure 5.3 The authors' method employs a Language-Interfaced Fine-Tuning (LIFT) framework alongside a Generative Pre-trained Transformer (GPT) to produce tokens that encapsulate the reaction conditions, including the synthesis procedure. The data on catalyst synthesis and testing are transformed into an embedding vector and linked to an objective function, such as C2 yield. They develop prompts that include multiple-choice options ("Multi," left panel) and single context completions ("Topk," right panel). In this framework, "multi" selects a single example ($k = 1$) from the list, whereas "topk" chooses two. The prompt begins with an instruction to the LLM regarding the desired outcome of the generated completion. Subsequently, formatted examples for in-context learning are constructed. Whereas "multi" generates five alternatives surrounding the label, "topk" employs only the label itself. Subsequently, the sequence to which the model is expected to respond is appended to both types of prompts. The two bottom panels display actual prompts crafted using the Multi (left panel) and Topk (right panel) templates. Each example demonstrates the input string to the LLM when selecting $k = 1$ for prompt creation. For "multi," the model produces all five options, but for "topk," the model generates only the numerical value of the prediction. Source: Ramos et al. (2023)/with permission of arXiv.

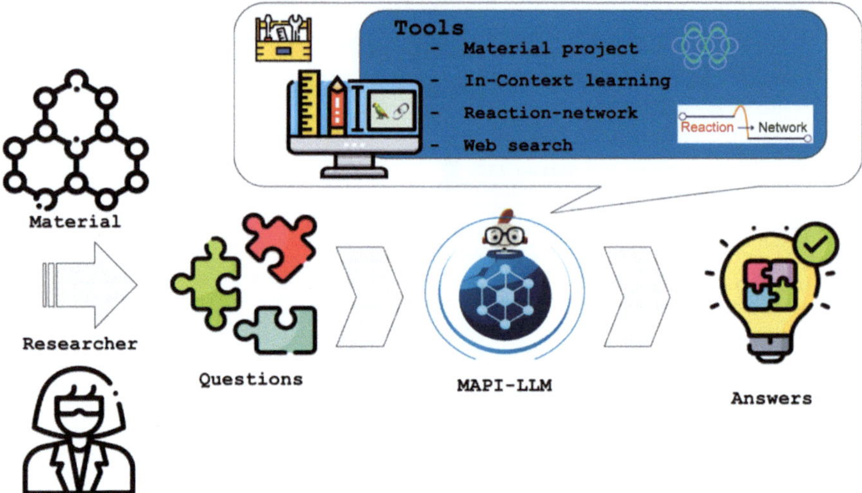

Figure 5.4 The operational process of the MAPI-LLM system. This system employs Language Models (LLMs) to interpret a researcher's query and strategically selects from an arsenal of tools – including the Materials Project API, the Reaction-Network package, and Google Search – following a step-by-step reasoning approach. Through this methodology, MAPI-LLM capably provides answers to inquiries concerning the stability of compounds, exemplified by the question "Is the material AnByCz stable?". Source: Jablonka et al. (2023a)/with permission of arXiv.

Another typical example of using LLM in material research could be automatic reference of external tools and knowledge, for which the MAPI-LLM is a typical one.[1] Figure 5.4 provides a schematic overview of this system. The MAPI-LLM framework represents a significant advancement in leveraging Language Models (LMs) for querying external knowledge bases, specifically in the realm of materials science. Developed by Mayk Caldas Ramos, Sam Cox, and Andrew White, MAPI-LLM is an innovative system designed to interface with the Materials Project API. This system is adept at transforming textual prompts into precise queries, thus facilitating answers to complex questions regarding material properties without the common pitfalls of hallucinations typically associated with automated systems.

For instance, when tasked with determining the stability of a material composition such as AnByCz, MAPI-LLM not only consults the Materials Project's extensive database for thermodynamic data but also, if necessary, initiates simulations for materials absent from the database. This level of interaction underscores the system's ability to bridge the gap between existing data repositories and computational simulations, ensuring a comprehensive coverage of material properties. Moreover, MAPI-LLM extends its capabilities to handle a variety of query types, from classification questions like the magnetism of Fe_2O_3 to regression problems concerning the band gap of $Mg(Fe_2O_3)_2$. This versatility is further exemplified in its ability to navigate through more intricate queries. For example, it can determine

---

1 Project can be referred at https://github.com/maykcaldas/MAPI_LLM, and tested online at https://huggingface.co/spaces/maykcaldas/MAPI_LLM.

the band gap of $Mn_{23}FeO_{32}$ by initially ascertaining whether it is metallic and proceeding accordingly if it is not. One of the most notable features of MAPI-LLM is its use of Inferred Contextual Learning (ICL). In scenarios where direct data on a material's property is unavailable, MAPI-LLM crafts a contextually rich ICL prompt using data from similar materials found in the Materials Project database. This approach allows the system to infer properties for unknown materials effectively, significantly reducing the risk of inaccuracies and hallucinations by leveraging relevant, external knowledge bases.

In summary, MAPI-LLM stands out as a pioneering solution that simplifies the interrogation of material properties through a seamless integration of LMs with external databases. By automating the translation of textual prompts into actionable queries and employing ICL to fill data voids, MAPI-LLM enhances the accuracy and reliability of electronic structure calculations and materials science research.

While the preceding examples have underscored the utilization of LLMs for predicting and generating materials, they essentially apply novel NLP or AI methods to tasks that, at their core, are not new. Tasks such as predicting material properties or reaction outcomes have been explored for decades. Essentially, these tasks treat AI as a tool, akin to conducting an experiment with AI as the assistant. However, with the emergence of emergent capabilities in LLMs, it's time to shift our focus toward leveraging them for more creative endeavors. Instead of forcing these models into roles they are not ideally suited for, such as straightforward prediction tasks that could limit their potential, we should consider deploying them in roles that maximize their unique capabilities. For instance, aiding in information gathering, analysis, and literature review could be areas where LLMs excel beyond mere predictive tasks. Therefore, this chapter advocates for the application of LLMs in more creative tasks, detailing our approach in utilizing these models not just as tools for prediction but also as partners in the creative process of scientific inquiry, particularly in ways that harness their ability to analyze and synthesize information from vast datasets.

Despite the diversity in focus and the emphasis on citation for content validity, the depth of LLM integration into scientific processes remains a subject of debate, underscoring the necessity for a more sophisticated fusion of LLMs within scientific workflows (Sanderson 2023). The benefits of LLMs stretch far beyond mere support in routine experimental tasks; they encapsulate the potential to forecast future research directions and critically assess scholarly contributions. This predictive ability, akin to conducting time-series analysis, coupled with the capacity to evaluate academic papers, fulfills the essential demand for tools that extend beyond operational assistance. Such models are instrumental in assisting researchers to navigate the vast expanse of literature, pinpointing innovative and impactful research ideas in the face of the prevalent challenge of over-publishing, which tends to dilute the quality of scholarly dialogue (Akbashev and Kalinin 2023). Additionally, the function of references in interlinking a wide array of publications underscores the progressive nature of scientific discovery, where new insights are built on the foundational works of predecessors. In this vein, LLMs act as digital catalysts for ideation, extracting and amalgamating ideas from millions of texts to spark new research avenues and interdisciplinary innovations. This attribute is indispensable for researchers striving to make significant contributions to their fields, requiring an elevated level of scholarly skill and insight to pioneer novel territories.

As the scholarly community navigates the dual hurdles of ensuring the originality and potential impact of research ideas, the integration of LLMs into the review process stands out as a promising strategy. Despite ongoing debates (Donker 2023), the application of AI tools in parsing the vast sea of publications can spotlight groundbreaking ideas (Kousha and Thelwall 2023), guaranteeing that they garner the recognition and attention they merit. Therefore, the following part of this chapter will focus on the application of LLM in the idea generation and novelty analysis, which can be concluded by using LLM to review the quality of publications and generate new research ideas.

This section sets the stage for an in-depth examination of how AI influences scientific exploration in two pivotal ways: nurturing the birth of original research concepts and critiquing scholarly articles for their innovative content. A notable development in this field is the introduction of a deep learning AI mentor, powered by a correlation-driven network of scholarly works known as ScholarNet. This AI mentor acts as a digital guide, offering researchers advice on possible research paths, assessing the originality of academic papers, and providing detailed recommendations. Through ScholarNet, which mirrors the structure of the famous ImageNet database used in computer vision, we can conduct a nuanced, large-scale evaluation of scholarly articles, surpassing the limitations of traditional keyword search techniques. The AI mentor's proficiency, demonstrated by its ability to forecast research trends in domains such as perovskite and MXene using tensor networks (TNs) and to derive extensive insights from abstracts, represents a stride toward employing AI in the conception of ideas and fostering cross-disciplinary breakthroughs.

Furthermore, the influence of AI on scholarly publishing is accentuated with the advent of the Spider Matrix. This novel platform enables a thorough examination of research papers and patents, judging them across various metrics, including originality, thoroughness, and influence, going beyond simple citation counts to provide a quick, fair appraisal based on the inherent merit of the work. The Spider Matrix's flexibility across different academic domains and its contribution to nurturing creative thought exemplify AI's transformative potential within the peer-review process and its capacity to ensure that inventive research is aptly acknowledged.

The synergy between these two AI applications – one for assessment, the other for idea generation – paints a picture of a future where scientific endeavors are not just expedited but also reinvented, leading to a more profound interaction with the growing body of scholarly work. By democratizing the evaluation of academic papers and endorsing a universal approach to scholarly investigation, platforms like Spider Matrix seek to enhance the academic realm's understanding and recognition of research, pushing toward a more collaborative and forward-thinking scientific milieu. The ensuing chapter will explore how AI, through mechanisms such as the AI mentor and Spider Matrix, can act as an impetus for conceiving new ideas and securing top-notch research findings, thus transforming the field of scientific research and academic publishing.

5.2 Fundamentals of LLM for Material Research: Database and Knowledge Base Construction

The core to acquire data is the careful collection of recent articles from RSS feeds, focusing on reputable, open access journals. We aim to receive a consistent daily stream of papers to categorize and merge into our database. These categories include Metadata, Citation links, and concise semantic summaries that highlight key elements such as the research question, theoretical approach, methods, findings, and conclusions. We also utilize public APIs like those from CrossRef, PubMed, and OpenAlex to gather essential details such as DOIs, titles, authors, and citations for journals chosen based on scimago and JCR rankings, specifically those published after 2006. This method provides us with a detailed and expansive article database that combines bibliometric information with in-depth semantic evaluation.

In developing an AI for assessing scholarly work, we've distilled human evaluative standards into an AI-comprehensible five-dimensional matrix, refining academic value assessment (referenced in (Superchi et al. 2019; Yoganandan and Vasan 2022)). Briefly:

Question: Evaluates the paper's innovation and relevance in posing research questions, gauging the inquiry's originality and exploration rationale.

Theoretical Approach: Assesses the theories' novelty and intellectual rigor, examining theoretical creativity and logical robustness.

Experimental Methodology: Analyzes the originality and thoroughness of experimental design and data collection methods, including replicability.

Effects and Impacts: Investigates the research outcomes' significance and originality, the evidence's strength, and their field contribution.

Conclusions and Contributions: Evaluates the insights and novel perspectives the research offers, and how the findings substantiate the conclusions.

This framework enables the AI to critique research articles comprehensively, reflecting the depth of traditional peer review. This process can be viewed in Figure 5.5.

To enhance our system, we utilize DOIs to access the full texts or abstracts of papers within copyright limits. We then convert this information into a structured format for in-depth analysis, primarily through word embedding tokens and feature maps. For papers with full-text access, we distill the content to about 1000 words, enabling LLMs to manage otherwise prohibitively long articles. We employ LangChain's ConversationChain method to address LLMs' text length constraints, breaking down articles into key elements for a conversational, sequential input into the LLM. This approach aids the model's comprehension and produces succinct summaries within a 500-word limit for each article section, allowing the LLM to analyze thoroughly. MatSciBert further refines our technique by extracting keywords with weighted importance, crafting focused summaries that highlight essential research points. This enhances the model's ability to pinpoint and underscore significant themes and methods, which aids in identifying similarities between papers and their connections, as demonstrated in Figure 5.5.

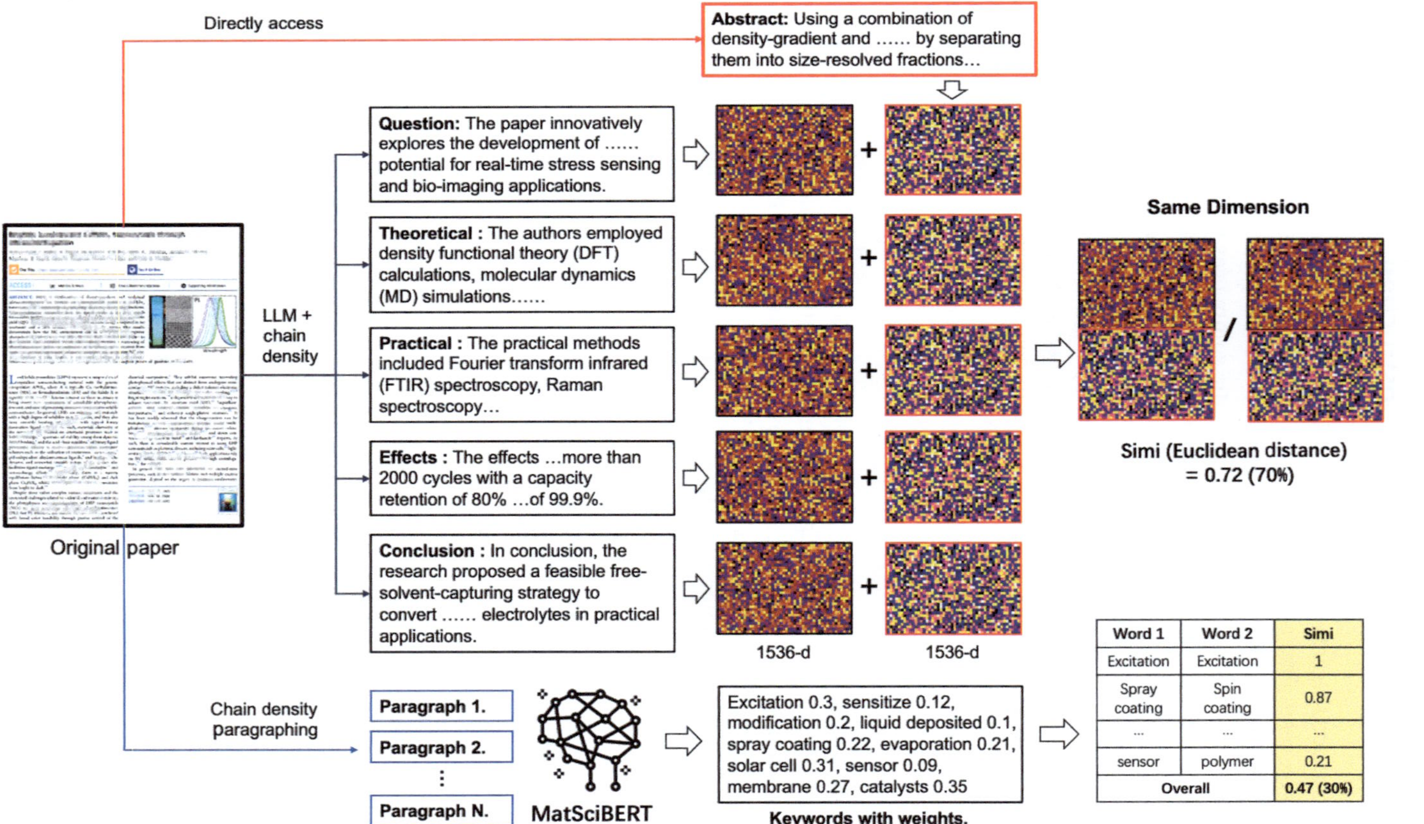

Word 1	Word 2	Simi
Excitation	Excitation	1
Spray coating	Spin coating	0.87
…	…	…
sensor	polymer	0.21
Overall		0.47 (30%)

Figure 5.5 The process for deriving features and computing similarities among scholarly papers. This involves converting the abstract into a 1536-dimensional vector and augmenting it with selective, sparse keywords created by MatSciBERT. The fusion of these dense and sparse data points creates a detailed profile, with varying color tones reflecting the degree of similarity, which assists in the accurate evaluation and differentiation of scientific articles. Source: Gupta et al. (2022)/Springer Nature/CC BY 4.0.

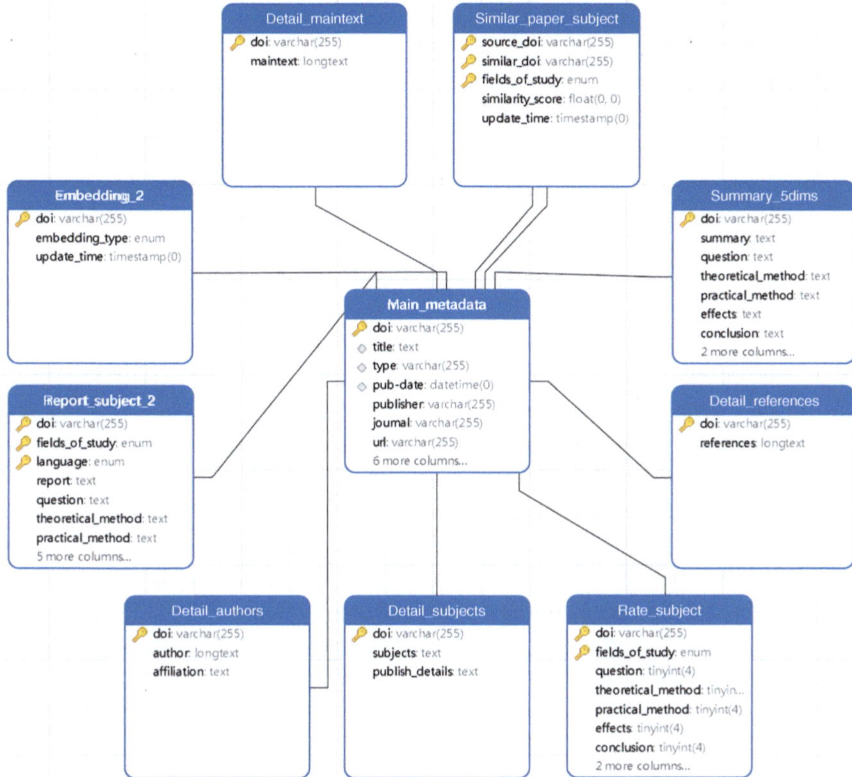

Figure 5.6 Database Schema for Academic Article Storage and Analysis. This diagram presents the structural design of a database system that stores detailed metadata, full texts, and summaries of scholarly articles. It outlines the interrelations between tables that hold article DOIs, metadata, author details, subject reports, embedding information, reference details, and similarity scores between papers, facilitating comprehensive data management and retrieval for academic evaluation and research trend analysis.

Our advanced pre-processing framework combines state-of-the-art data gathering with analytical techniques, preparing the ground for AI-based article evaluation and a deeper understanding of academic content. For our database, we integrate a robust system on a custom cloud platform, combining MySQL for structured data and Pinecone for managing vectorized data. This infrastructure supports the organization and retrieval of our extensive collection of around 12 million articles. MySQL organizes structured data like metadata and article overviews, covering a wide range of publishers. The structure can be seen in Figure 5.6. Pinecone deals with vectorized summaries, allowing advanced semantic search capabilities for content-based querying, thus enhancing our AI supervisor's recommendation and analysis efficiency. This setup strengthens our platform's ability to process and analyze the academic dataset's breadth.

5.3 Evaluation: Spider Matrix

We now present the Spider Matrix, a sophisticated tool designed to evaluate the academic merit of recent scholarly publications. The Spider Matrix operates through a two-step procedure: it begins by pinpointing articles with related themes and then appraises new articles against this established framework. Utilizing RSS feeds, the system gathers daily updates of publications and categorizes them into Metadata, Reference links, and an elaborate semantic synopsis that encapsulates each article's research questions, theoretical framework, methods, results, and conclusions (Figure 5.7).

For every one of these categories, a benchmark group of similar articles is compiled and abstracted, providing a reference by which the originality and scholarly achievements of new articles are gauged. The system meticulously assesses the extent to which a new article aligns with or diverges from the existing body of work in its field, with a particular emphasis on its originality and scholarly impact. Consequently, articles are assigned a score indicative of their scholarly worth and potential impact in the academic community. The Spider Matrix thus offers a structured method for determining the quality of academic papers, enhancing insight into their scholarly contributions and their place in the wider academic dialogue.

In the process of deconstructing the information contained in academic papers, the abstract serves as a concise summary, capturing the key points of the study. This condensed information is then encoded into a numerical form, specifically a 1536-dimensional vector, through OpenAI's text-embedding-ada-002 model. This conversion facilitates the abstract's integration into our analytical framework for easier comparison by our system. Subsequently, the LLM utilizes its advanced linguistic analysis capabilities to break down the article into five distinct segments (problem, theoretical approach, experimental methodology, results, and conclusion), creating tailored summaries for each section. During this phase, tables and figure captions, along with images processed by GPT-4 Vision, are also converted into text to ensure a thorough narrative description that encapsulates all vital data from the paper. For content that exceeds the LLM's token limit, we employ the Chain of Thought approach, limiting summaries to 500 words per dimension. We also assess various LLMs' proficiency in summarizing scholarly content, with those details provided in the supplementary results (Du et al. 2021; Jiang et al. 2023; Zeng et al. 2022; Zhang et al. 2023). These summaries are then transformed into 1536-dimensional vectors, similar to the initial abstract encoding, and combined to form a detailed 3072-dimensional feature matrix that encompasses both broad and detailed perspectives of the paper.

To refine the analysis of articles laden with complex terminologies, particularly in fields such as Material Science and Chemistry, we integrate the MatSciBERT model. This model is adept at pinpointing field-specific terms and producing a list of pertinent keywords. These keywords are essential to the article's feature matrix, addressing the general embeddings' shortfall in capturing the depth of technical language. In the final feature matrix, used to compare and evaluate articles, these domain-specific keywords make up 30% of the matrix, while the remaining portion

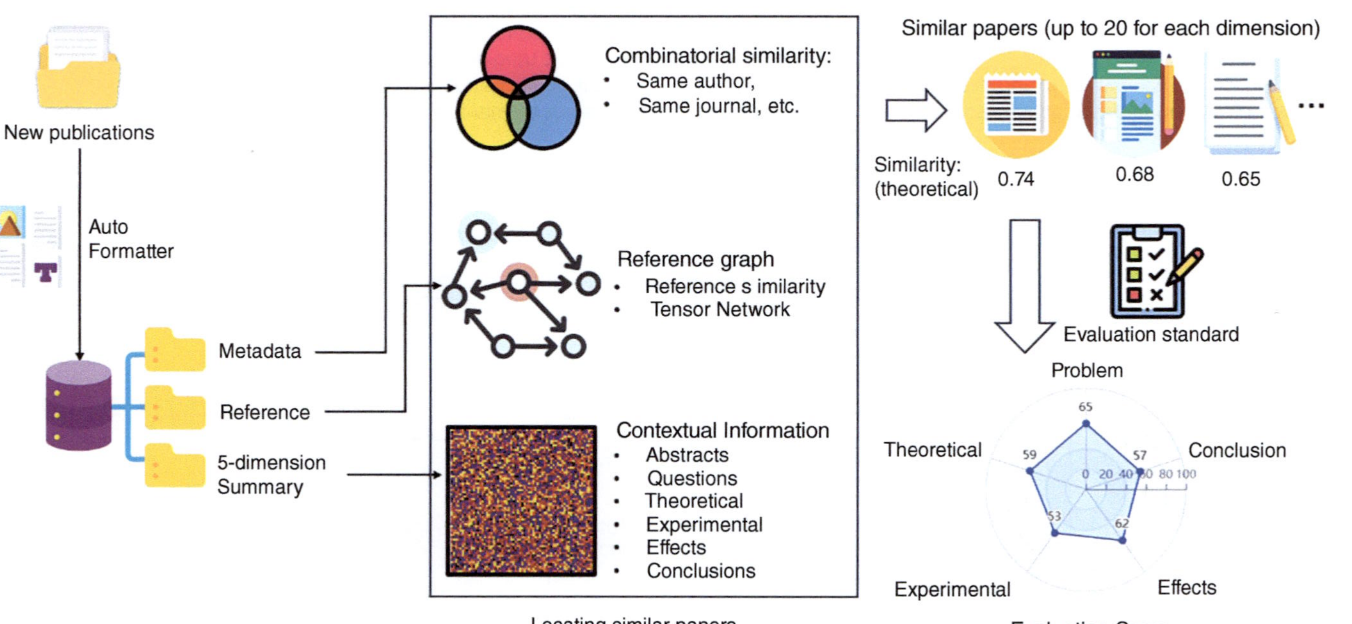

Figure 5.7 Overview of Spider Matrix, a two-stage system for identifying and evaluating new scholarly articles. The process involves sorting through Metadata, Reference links, and semantic classification to perform an in-depth assessment. The final evaluation is carried out across five separate dimensions for a complete analysis.

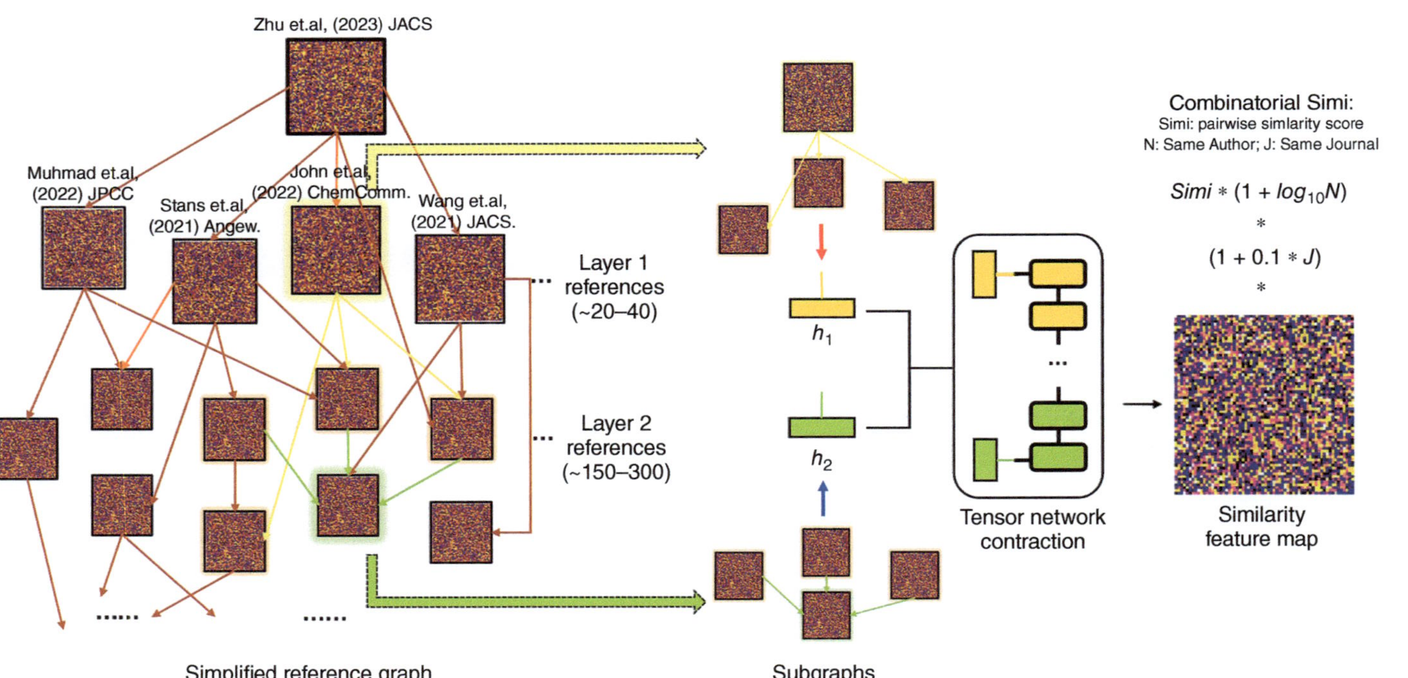

Figure 5.8 The creation of a detailed feature matrix to evaluate the similarity among scholarly articles, a process elaborated in Figure 5.5. It starts with transforming abstracts into vectors which are then integrated with data from the citation network through tensor graph networks. The scalar similarity metrics are enhanced through tensor contraction to include both the content of the articles and their citation contexts. A scaling mechanism corrects for potential biases due to common authorship or journal of publication, as shown in the upper right. This results in an advanced similarity matrix that captures the nuanced connections between articles, both in terms of their content and their place within the citation network.

consists of the previously mentioned contextual embeddings, as depicted on the right side of Figure 5.5. This combination not only allows for a more profound understanding of the articles but also significantly contributes to the nuanced assessment based on article comparisons.

To construct a nuanced similarity feature matrix, relying solely on basic similarity measures may not suffice, as these may overlook the intricate web of citations and interconnections between scholars and their work. To address this, our approach incorporates a tensor graph network as detailed in Figure 5.8, merging the core concepts derived from article abstracts with citation data to deepen our grasp of the articles' significance and interrelations. Our system simplifies by representing articles primarily through their abstracts within this network, balancing efficiency with the retention of essential information. Through tensor contraction processes (Huang et al. 2021; Ran et al. 2020), which will be further explained in later chapters, we synthesize this data into a comprehensive similarity matrix that's part of Scholar-Net. This matrix goes beyond content resemblance, factoring in citation linkages to provide a holistic similarity perspective.

We apply adjustments to the similarity scores to neutralize biases from common authorship or publication in the same journal, ensuring a more equitable comparison, as demonstrated on the right of Figure 5.8. The resulting matrix is both balanced and intricate, offering an improved tool for evaluating article quality and relevance. Furthermore, our system identifies roughly 10 comparable papers to a given article (adjustable between 3 and 99 as necessary). To gauge the novelty of the research, we compare against papers published at least three months prior to the article under review, setting an appropriate benchmark for assessing originality.

To establish benchmarks for new research, we designed a detailed questionnaire for evaluators, capturing basic information, full texts, summaries, and identified similar works. The assessment spans various dimensions – problem, methodologies, results, and conclusions – each probed with targeted questions scored from 1 to 10, as exemplified in Figure 5.9a. This approach evaluates the research's originality and contribution, providing both scores and qualitative insights. We calculate the average scores for each dimension using the two highest-rated questions and the top score from related articles, emphasizing the research's strongest points and translating the scores into percentile ranks. A LLM's evaluation accuracy is honed through a questionnaire analysis based on 200 reviews by experts, fine-tuning it to the scoring standards of Chemistry and Material Science, as depicted in Figure 5.9a.

Figure 5.9b outlines our novel evaluation method, serving as both a scoring guide and an interpretative tool. At its core is the innovation score, gauging the article's distinctiveness against a vast database from leading journals. This score is a gauge of innovation, assessing the article's divergence from prior studies, particularly in the "effects" dimension, as shown in the right of Figure 5.9b. Reports supplementing each dimension offer textual analysis, reflecting the paper's originality and academic quality. Papers with high similarity scores receive lower innovation evaluations, while those with less resemblance to prior work score higher. Highly innovative but scarcely cited papers are scrutinized further across theoretical, experimental, and effort dimensions to confirm their validity. This

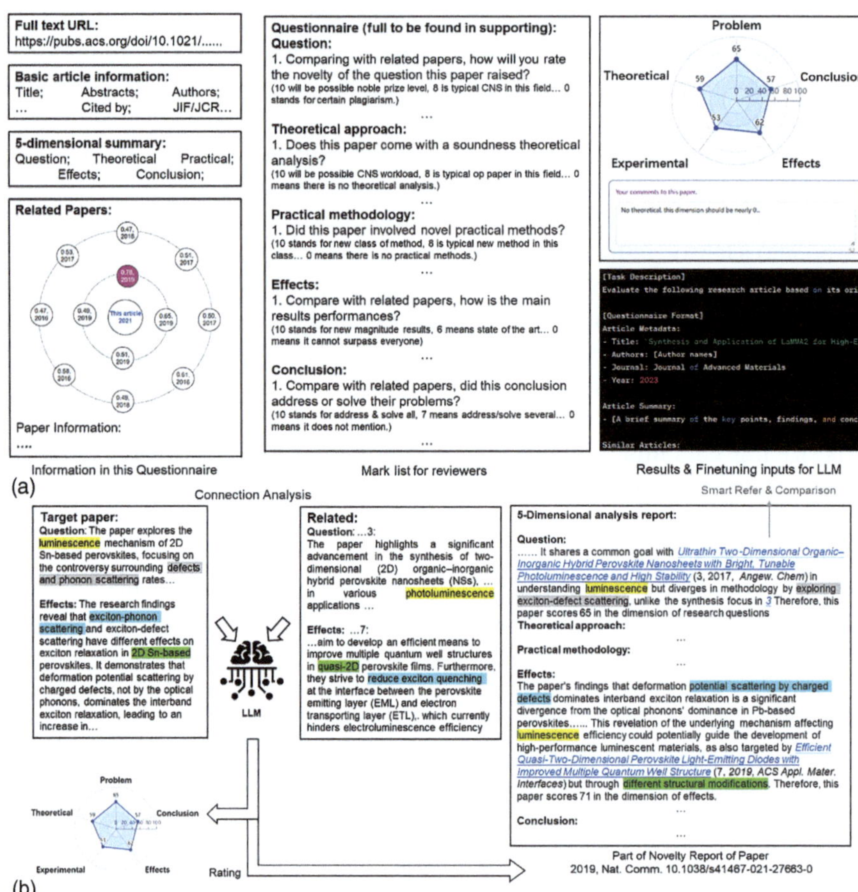

Figure 5.9 The process of honing a Language Model (LLM) for the specialized task of academic paper scoring, complete with an evaluative report. (a) A detailed questionnaire designed for reviewers to assess scholarly articles, including key metadata, content synopses, and likenesses to related studies. This questionnaire applies a 1–10 rating scale across various dimensions, such as originality and research methods, and benchmarks articles against prior research. The top scores across these dimensions are averaged for the final evaluation, a step that contributes to refining a LLM for domain-specific scoring accuracy. (b) The intricate steps involved in forming the innovation-level analysis and its accompanying report. This includes a quantitative innovation score derived from a multi-faceted analysis, where the article's similarity to existing literature plays a crucial role in score determination. Keywords from the inputs and outputs are color-coded to show correlations, and citations are included in the corresponding sections of the five-dimensional report. The system currently supports only open access publications.

balanced method delivers a comprehensive assessment of the true innovative value of scientific research.

The transition from paper evaluation to innovation generation is depicted in associated illustrations, delineating a sophisticated trajectory that begins with an in-depth analysis of scholarly articles. By amalgamating features, keywords, and citation data, we establish a structure akin to a knowledge graph, which

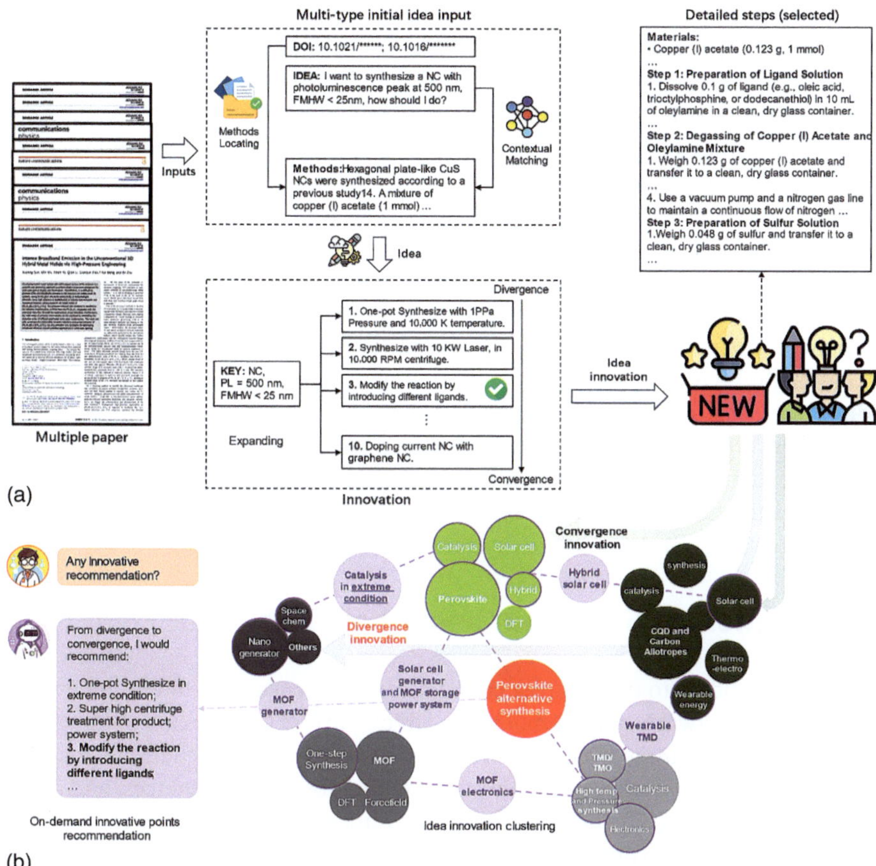

Figure 5.10 The process from article assessment to the generation of innovative ideas. In part (a), the diagram shows the transformation of multiple paper inputs into points of idea innovation, with an instance provided for integrating different ligands into a reaction process. Part (b) delves into the mechanisms of generating novel ideas based on the research concepts of each paper. The Spider Matrix is utilized to discover potential convergences among the articles, presenting the outcomes interactively.

facilitates the identification of novel connections. Within this structure, we recognize two key innovation avenues: "Convergence innovation" involves incremental improvements within established concept clusters, typically generating practical yet incremental advancements. In contrast, "Divergence innovation" encourages bold, cross-disciplinary ideation that spans disparate concept domains, potentially yielding highly original ideas with greater risk.

Figure 5.10a outlines the steps of innovation generation. The Spider Matrix system assimilates various inputs such as scientific papers and concepts, extracting methods sections to identify key procedures. It conducts semantic analyses to locate related works and synthesizes this information into a structured methods outline, catalyzing the innovation process. A human evaluator then selects the most promising idea, balancing inventiveness with feasibility, with the third idea from the innovation list often chosen for further development. Delving into the mechanics of idea generation in Figure 5.10b, the system navigates the knowledge graph-like framework to compile a comprehensive, actionable list of steps from existing research, offering researchers a practical guide. Initially, these steps may be manually implemented, but as the conversation on laboratory automation progresses, they could be converted into code for automated laboratory systems, as recent studies (Li et al. 2020a,b; Xu et al. 2023) have explored.

Our approach to innovation assessment aims to embolden the exploration of uncharted ideas, pushing researchers beyond conventional convergence to discover pioneering, transformative concepts. Our framework is designed to preferentially spotlight divergent proposals. While these require human discernment for practicality and might be further refined by computational experimentation, the emphasis is on broadening the scope of scientific inquiry and discovery. The strategy is not to forgo convergence but to proactively pursue divergence, encouraging significant leaps in creativity that have the potential to alter scientific fields fundamentally. This way, our evaluation framework assesses and simultaneously inspires innovation, acting as a catalyst for the conception of trailblazing ideas.

5.4 Ideation: AI Supervisor and ScholarNet

We initiate a study into the complex connections within academic literature by creating a citation network called ScholarNet, which mirrors the function of ImageNet in the realm of computer vision. ScholarNet provides a comprehensive base for our AI Supervisor, containing a wealth of academic abstracts, references, and other scholarly information. At the heart of the AI Supervisor are two key elements: a TN-driven citation network and a sophisticated feature extraction tool that employs few-shot learning. This system stands out for its capacity: (1) to identify research trends in materials science, especially for substances like perovskite and MXene; (2) to perform in-depth analysis of abstracts for enhanced article review; and (3) to continuously update and track changes within the ScholarNet database. This approach allows for a more detailed and expansive analysis of academic work than traditional keyword searches.

The AI Supervisor plays a crucial role in impartially assessing a paper's originality and its publication's necessity, a topic explored above as part of the Spider Matrix. Moving from evaluation to ideation, traditional analysis methods typically fall under concept-focused173 or citation-driven174 categories. Our system combines these methods, refining concept identification to suit material science and chemistry better. Beyond simple keyword extraction detailed above, Figure 5.11a demonstrates an expanded approach to derive a broader range of keywords, capturing the entire scope of research, particularly for abstracts and full papers in these fields. This enhanced method facilitates the extraction of key standardized data, integral to the feature map, via a few-shot learning technique illustrated in Figure 5.11b. Our model is adept at discerning essential elements from unpublished texts and autonomously gauging their innovativeness. For published papers, it also generates a novelty score by assessing citation linkages within ScholarNet.

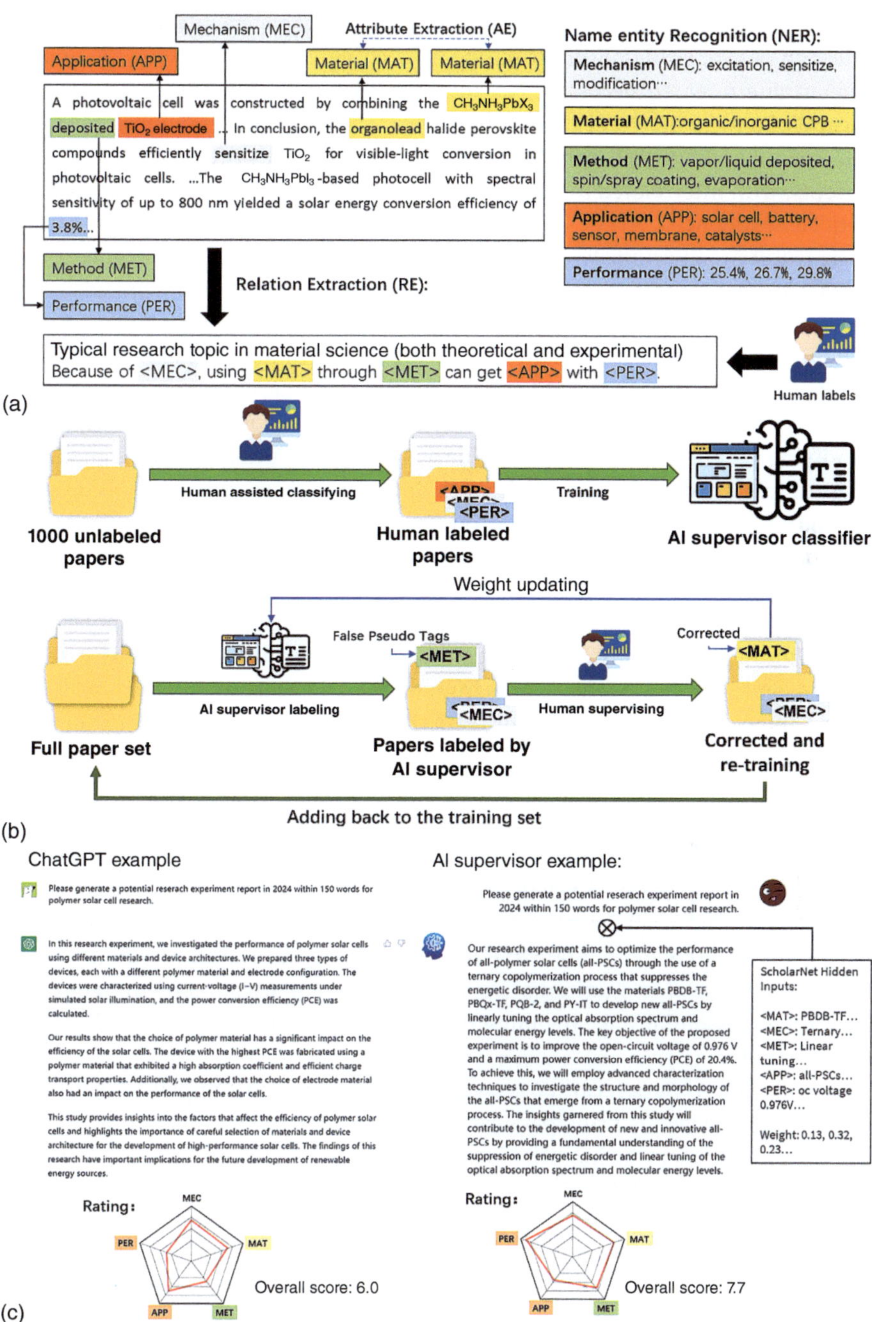

Figure 5.11 The novelty rating methods and results. (a) The material science paper key information extraction method. (b) The functional diagram of data labeling method by AI Supervisor. (c) The comparison results of ChatGPT and AI Supervisor research guiding function (to predict a potential research experiment report in 2024). The main difference is that AI Supervisor contains a hidden input generated by ScholarNet, while ChatGPT does not. In result, comparing with few-keywords inputs, AI Supervisor suggested keywords resulted in a 28.3% higher score in this case (and average 18.7% in 10,000 trials). Source: Xu et al. (2023)American Chemical Society.

Figure 5.11c presents a tiered quality categorization of papers, utilizing ScholarNet to discern pivotal academic contributions and a five-dimensional keyword analysis to determine research necessity. An algorithm that considers the current interest ("heat") of a topic area and the interrelation ("co-occurrence") of ideas across the field calculates necessity scores. This scoring reflects both the focus and scope of concepts, with interrelated topics like perovskite and solar cells contributing to the overall assessment.

The feature map arranges up to 30 significant keywords in five categories, each with a relevance score shown in Figure 5.11c. These categories create linked clusters that enhance ScholarNet's depth of analysis. Integrating key ScholarNet insights with the Language Learning Model (LLM) efficiently addresses major paper topics. As illustrated on the right of Figure 5.11c, the AI Supervisor boosts manuscript quality by pinpointing critical focus areas, uncovering neglected subjects, and identifying potential for cross-disciplinary synergy. By utilizing this comparative analysis, researchers can hone their work, leveraging these cross-field connections and addressing identified gaps. The AI Supervisor thus becomes an indispensable asset for materials science researchers to elevate the caliber and significance of their scholarly output.

The AI Supervisor utilizes keywords and historical insights from ScholarNet to project future publications, as indicated in Figure 5.12(a, with predictions spanning 1–2, 3–4, and 5–6 years ahead. Works from these periods are highlighted in red, displaying their level of similarity and titles. However, the AI Supervisor's ability to conceive brand new concepts or domains weakens over time due to its dataset only covering until 2015, making its longer-term predictions less dependable. For anticipated scientific developments in 2024, as shown in Figure 5.12b, the AI Supervisor identified two noteworthy topics concerning MAPbBr3. The first topic (Marked as Selected Result 1 in Figure 5.12b) is the synthesis of a superconductor under extreme conditions, a notable yet daunting task. The second topic (Marked as Selected Result 2 in Figure 5.12b) is more attainable but less likely to lead to extraordinary breakthroughs because of its traditional approach. Balancing feasibility with pioneering innovation is essential since the pursuit of revolutionary ideas, although difficult, is highly prized.

Typically, the AI Supervisor projects either innovative but complex or feasible yet unremarkable ideas. Its principal goal is to reveal unexplored research niches that could steer new scientific inquiries. This tool's proficiency was evaluated in various fields, including materials science and biomedicine, as illustrated in Figure 5.12c. Based on specific keywords, it produced predictions, with their validity gauged against actual research themes. Over time, the prediction accuracy tends to wane, a consequence of the data being capped in 2015. Forecasts in materials science were more on point compared to those in CRISPR technology, mirroring their representation in the dataset. The AI Supervisor has notably enhanced the pertinence of research keywords, especially when merged with generative LLMs, registering a significant boost in relevance, with a 28.3% improvement in certain cases and an average enhancement of 18.7% across multiple evaluations. While some recommendations might not be feasible, the AI Supervisor encourages creative thinking,

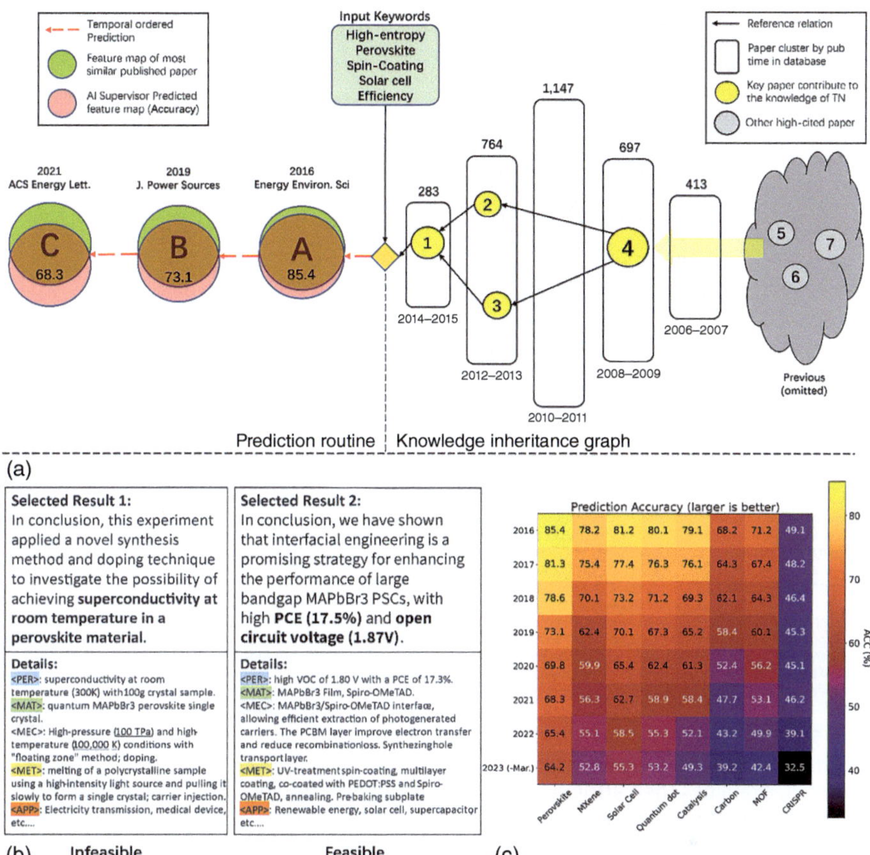

Figure 5.12 AI Supervisor's prediction routine using a partial pre-2015 database. (a) ScholarNet's functional diagram for knowledge inheritance and future prediction (top center). Left: Prediction accuracy compared to the most similar published paper. Right: ScholarNet knowledge inheritance graph with key papers from 10 years marked yellow. Older publications were omitted. Marked papers A–C and 1–7 correspond to Figure 5.13 captions. (b) Two AI Supervisor 2024 potential experiment report examples: the left discusses challenging superconductivity under extreme conditions, while the right one on solar cells is more feasible. (c) Prediction accuracy comparison across five categories and time. Accuracy drops quickly and varies between categories due to the pre-2015 training database, with most root papers being perovskite related. Source: Reproduced with permission from Xu et al. (2023).

providing strategic insights for navigating the research field. It also aids in various tasks, such as calculating novelty scores for authors and detecting key papers within each discipline using ScholarNet. In conjunction with ScholarNet, it contributes significantly to training LLMs to tackle problems in materials science. Nonetheless, a deep comprehension of the subject matter, supplemented by hands-on learning, is indispensable. As the database expands, it deepens our understanding of the evolution of materials science, with well-categorized data sharpening our knowledge of research approaches. Although platforms like Galactica offer comparable functions, their susceptibility to errors limits their practical application.

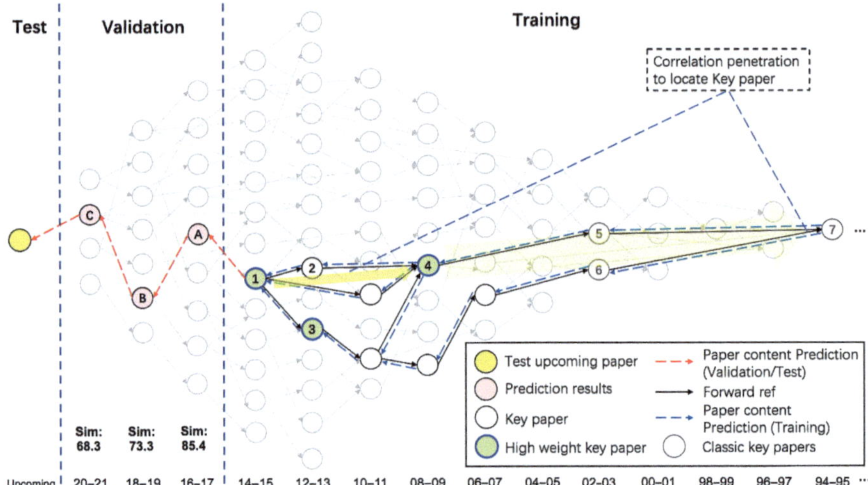

A. Entropic stabilization of mixed A-cation ABX3 metal halide perovskites for high performance perovskite solar cells
B. One-step-spin-coating route for homogeneous perovskite/pyrrole-C60 fullerene bulk heterojunction for high performance solar cells
C. Temperature-Dependent Ionic Conductivity and Properties of Iodine-Related Defects in Metal Halide Perovskites

1. Compositional engineering of perovskite materials for high-performance solar cells.
2. Efficient organometal trihalide perovskite planar-heterojunction solar cells on flexible polymer substrates.
3. Efficient hybrid solar cells based on meso-superstructured organometal halide perovskites.
4. Organometal halide perovskites as visible-light sensitizers for photovoltaic cells.
5. Comparative study on the excitons in lead-halide-based perovskite-type crystals CH3NH3PbBr3 CH3NH3PbI3.
6. Optical properties of CH3NH3PbX3 (X = halogen) and their mixed-halide crystals.
7. Optical properties of PbI-based perovskite structures.

Figure 5.13 A comprehensive visualization of knowledge progression via ScholarNet citations. Using targeted keywords, the system traces pivotal historical publications within ScholarNet, examines their attributes, and integrates these findings into new tests and validations. Notable referenced works include: A: (Energy Environ. Sci., 2016, 9, 656–662); B: (J. Power Sources., 2019, 45, 27–34); C: (ACS Energy Lett., 2022, 7, 310–319); 1: (Nature, 2015, 517, 476–480); 2: (Nat Commun., 2013, 4, 2761); 3: (Science, 2012, 338, 643–647); 4: (J. Am. Chem. Soc., 2009, 131, 6050–6051); 5: (Solid State Commun., 2003, 127, 619–23); 6: (J. Mater. Sci., 2002, 37, 3585–3587); 7: (J. Lumin., 1994, 60–61, 269–274). Source: Xu et al. (2023)/American Chemical Society.

5.5 Results and Discussion

We performed objective results analysis, where the results have been illustrated in Tables 5.1 and 5.2. Objectively evaluating the results from the AI Supervisor, our analysis finds that the model exhibits notable prediction accuracy. This accuracy is quantified using the average cosine similarity between the vector representations of predicted and actual keywords, calculated as follows:

$$\text{Acc} = \frac{1}{n}\sum_{i=1}^{n}\frac{P_n \cdot A_n}{|P_n||A_n|}$$

However, a critical observation is that two keywords with a high degree of vector similarity might have distinct meanings in certain contexts. For instance, in materials science, "electrolysis" and "electromagnetism" may have similar vector values but are conceptually different. On the other hand, "gas sensors" and "solar cells" may not appear directly related but often share fundamental techniques and principles.

Table 5.1 The accuracy in reference paper prediction from 2016 to 2023.

Year	Cosine Acc	Top-10 Acc
2016	0.93	0.37
2017	0.92	0.31
2018	0.89	0.23
2019	0.90	0.15
2020	0.88	0.10
2021	0.85	0.07
2022	0.83	0.06
2023	0.81	0.05

Table 5.2 Predicts most popular research hotspot from 2016 to 2023, based on the publication database of 2015.

Year	Predicted most popular research hotspot	
2016	**Perovskite stability improvement**	**MXene synthesis and property**
2017	Perovskite LED	MXene energy storage
2018	Perovskite lasers	MXene catalysis
2019	Perovskite transistors	MXene nanomaterials
2020	Perovskite memory devices	MXene water treatment
2021	Perovskite energy storage	MXene stability
2022	Perovskite optoelectronic devices	MXene cost-effective synthesis
2023	Advanced synthesis of perovskite	MXene flexible devices

To refine our accuracy assessment, we implement a top-k accuracy metric. This metric evaluates how often the actual keywords appear among the model's top 20 predictions, thereby offering a more precise gauge of the model's performance in predicting research topics.

$$\text{Acc}_{\text{TopK}} = \frac{1}{n}\sum_{i=1}^{n} \frac{TP_i}{TP_i + FP_i}$$

This additional measure helps in assessing the exactness of the model's keyword predictions, enhancing the evaluation of how well the predicted topics correspond with the genuine research themes.

The primary objective of forecasting future publications from preceding ones is detailed in Table 5.1.

For instance, predictions regarding key areas of interest in perovskite solar cell research from 2016 to 2019, utilizing data up to 2015, are outlined in this table. Looking back from 2023, the forecasts for 2016 and 2017 appear to be well founded, though the value of 2018's predictions diminishes due to similarities with LED research. Predictions for 2019 and 2020 were found to be overly optimistic, as areas like perovskite transistors and memory devices have not seen substantial research development.

5.6 Conclusion

The integration of AI tools like the Spider Matrix and AI Supervisor is reshaping academic publishing and research innovation. The Spider Matrix excels in multi-faceted evaluations of academic papers, surpassing traditional metrics by focusing on the quality and relevance of research rather than publication volume. Its adaptability to different disciplines and patents demonstrates its wide-ranging utility. The AI Supervisor, with its advanced analysis and predictive capabilities, particularly in materials science, has become an essential aid for researchers, helping them trace scientific progress and forecast upcoming trends with a high degree of accuracy.

Together, these AI systems represent a transformative step in scholarly evaluations, bridging the gap from critical assessment to creative ideation. Their ability to process a broad spectrum of research and patents positions them as pivotal tools for modernizing how academic merit and intellectual property are appraised. Promising a fair and rigorous approach to research evaluation, these tools stand to cultivate an informed, innovative research community ready to embrace the next wave of scientific breakthroughs. This succinct conclusion highlights their synergistic impact on fostering progress and ingenuity in the academic world. We believe that the application of LLMs in Material Science should extend beyond simple knowledge repositories and predictive tasks to advanced applications that generate new knowledge and critically assess existing information.

References

Akbashev, A.R. and Kalinin, S.V. (2023). Tackling overpublishing by moving to open-ended papers. *Nature Materials* 22 (3): 270–271. https://doi.org/10.1038/s41563-023-01489-1.

Brown, T., Mann, B., Ryder, N. et al. (2020). Language models are few-shot learners. *Advances in Neural Information Processing Systems* 33: 1877–1901.

Dinh, T., Zeng, Y., Zhang, R. et al. (2022). Lift: language-interfaced fine-tuning for non-language machine learning tasks. *Advances in Neural Information Processing Systems* 35: 11763–11784.

Donker, T. (2023). The dangers of using large language models for peer review. *The Lancet Infectious Diseases* 23 (7): 781. https://doi.org/10.1016/S1473-3099(23)00290-6.

Du, Z., Qian, Y., Liu, X. et al. (2021). *Glm: General Language Model Pretraining with Autoregressive Blank Infilling*. arXiv preprint arXiv:2103.10360.

Goldberg, Y. and Levy, O. (2014). *word2vec Explained: Deriving Mikolov et al.'s Negative-Sampling Word-Embedding Method*. arXiv preprint arXiv:1402.3722.

Gupta, T., Zaki, M., Krishnan, N.M.A., and Mausam. (2022). MatSciBERT: a materials domain language model for text mining and information extraction. *npj Computational Materials* 8 (1): 102. https://doi.org/10.1038/s41524-022-00784-w.

Hong, Z., Ajith, A., Pauloski, G. et al. (2022). *Scholarbert: Bigger is not Always Better. arXiv preprint* arXiv:2205.11342.

Huang, C., Zhang, F., Newman, M. et al. (2021). Efficient parallelization of tensor network contraction for simulating quantum computation. *Nature Computational Science* 1 (9): 578–587. https://doi.org/10.1038/s43588-021-00119-7.

Inciteful (2023). Using Citations to Explore Academic Literature. Retrieved from https://inciteful.xyz/

Jablonka, K.M., Ai, Q., Al-Feghali, A. et al. (2023a). 14 examples of how LLMs can transform materials science and chemistry: a reflection on a large language model hackathon. *Digital Discovery* 2 (5): 1233–1250. https://doi.org/10.1039/D3DD00113J.

Jablonka, K. M., Schwaller, P., Ortega-Guerrero, A., and Smit, B. (2023b). Is GPT-3 all you need for low-data discovery in chemistry? *ChemRxiv.* doi: 10.26434/chemrxiv-2023-fw8n4.

Jiang, A.Q., Sablayrolles, A., Mensch, A. et al. (2023). *Mistral 7B. arXiv preprint* arXiv:2310.06825.

Kousha, K., and Thelwall, M. (2023). *Artificial Intelligence to Support Publishing and Peer Review: A Summary and Review.* Learned Publishing, https://doi.org/10.1002/leap.1570

Krenn, M., Ai, Q., Barthel, S. et al. (2022). SELFIES and the future of molecular string representations. *Patterns* 3 (10): 100588.

Krenn, M., Häse, F., Nigam, A. et al. (2020). Self-referencing embedded strings (SELFIES): a 100% robust molecular string representation. *Machine Learning: Science and Technology* 1 (4): 045024.

Li, J., Li, J., Liu, R. et al. (2020a). Autonomous discovery of optically active chiral inorganic perovskite nanocrystals through an intelligent cloud lab. *Nature Communications* 11 (1): 2046.

Li, J., Tu, Y., Liu, R. et al. (2020b). Toward "on-demand" materials synthesis and scientific discovery through intelligent robots. *Advanced Science* 7 (7): 1901957.

Li, L.H., Yatskar, M., Yin, D. et al. (2019). *Visualbert: A Simple and Performant Baseline for Vision and Language.* arXiv preprint arXiv:1908.03557.

Lu, J., Batra, D., Parikh, D., and Lee, S. (2019). Vilbert: pretraining task-agnostic visiolinguistic representations for vision-and-language tasks. *Advances in Neural Information Processing Systems* 32. https://proceedings.neurips.cc/paper_files/paper/2019/hash/c74d97b01eae257e44aa9d5bade97baf-Abstract.html.

Minaee, S., Mikolov, T., Nikzad, N. et al. (2024). *Large Language Models: A Survey.* arXiv preprint arXiv:2402.06196.

Nicholson, J.M., Mordaunt, M., Lopez, P. et al. (2021). scite: a smart citation index that displays the context of citations and classifies their intent using deep learning. *Quantitative Science Studies* 2 (3): 882–898. https://doi.org/10.1162/qss_a_00146.

Olivetti, E.A., Cole, J.M., Kim, E. et al. (2020). Data-driven materials research enabled by natural language processing and information extraction. *Applied Physics Reviews* 7 (4): 041317. https://doi.org/10.1063/5.0021106.

Ought (2023). Elicit: The AI Research Assistant. Retrieved from https://elicit.org/.

PBC, A. (2023). Introducing Claude. Retrieved from https://www.anthropic.com/index/introducing-claude

Radford, A., Kim, J. W., Hallacy, C., et al. (2021). *Learning Transferable Visual Models from Natural Language Supervision*. Paper presented at the International Conference on Machine Learning.

Ramos, M.C., Michtavy, S.S., Porosoff, M.D., and White, A.D. (2023). *Bayesian Optimization of Catalysts With In-context Learning*. arXiv preprint arXiv:2304.05341.

Ran, S.-J., Tirrito, E., Peng, C. et al. (2020). *Tensor Network Contractions: Methods and Applications to Quantum Many-Body Systems*. Springer Nature.

Sanderson, K. (2023). AI science search engines are exploding in number — are they any good? *Nature* https://doi.org/10.1038/d41586-023-01273-w.

scite. (2023). scite: see how research has been cited. Retrieved from https://scite.ai/home.

Selva Birunda, S. and Kanniga Devi, R. (2021). A review on word embedding techniques for text classification. *Innovative Data Communication Technologies and Application: Proceedings of ICIDCA 2020*, 267-281.

Superchi, C., González, J.A., Solà, I. et al. (2019). Tools used to assess the quality of peer review reports: a methodological systematic review. *BMC Medical Research Methodology* 19 (1): 48:https://doi.org/10.1186/s12874-019-0688-x.

Xu, Y., Ye, S., and Zhu, X. (2023). The ScholarNet and artificial intelligence (AI) supervisor in material science research. *Journal of Physical Chemistry Letters* 14 (36): 7981–7991. https://doi.org/10.1021/acs.jpclett.3c01668.

Yoganandan, G. and Vasan, M. (2022). Evaluating the quality of scientific research papers in entrepreneurship. *Quality & Quantity* 56 (5): 3013–3027. https://doi.org/10.1007/s11135-021-01254-z.

Zeng, A., Liu, X., Du, Z. et al. (2022). *Glm-130b: An Open Bilingual Pre-trained Model*. arXiv preprint arXiv:2210.02414.

Zhang, P., Wang, X.D.B., Cao, Y. et al. (2023). *Internlm-xcomposer: A Vision-Language Large Model for Advanced Text-image Comprehension and Composition*. arXiv preprint arXiv:2309.15112.

6

Toward a Blockchain-Powered Anti-Counterfeiting Experimental Data System in an Autonomous Laboratory

In the rapidly evolving landscape of scientific research, the integrity and authenticity of experimental data have become paramount, especially in the fields of chemistry and materials science. The advent of autonomous laboratories, powered by artificial intelligence (AI) and robotics, heralds a new era of experimental synthesis, promising unprecedented efficiency and precision. However, this technological renaissance also brings forth challenges, particularly in safeguarding the data against counterfeiting and ensuring its veracity. This chapter delves into the innovative solution offered by blockchain technology, a decentralized digital ledger known for its robust security, transparency, and immutability features. This chapter aims to explore how blockchain can be seamlessly integrated into autonomous laboratory environments to create a foolproof system for recording, storing, and verifying experimental data. By establishing a blockchain-powered anti-counterfeiting experimental data system, we can not only protect the integrity of scientific research but also foster a culture of trust and collaboration within the scientific community. Through an examination of blockchain fundamentals, real-world applications, and potential challenges, this chapter sets the stage for a future where scientific experimentation is both secure and transparent, paving the way for groundbreaking discoveries in materials science.

6.1 Blockchain Technology

Blockchain is a distributed ledger technology that records information through a series of blocks linked using cryptographic hashes. Each block contains the cryptographic hash of the previous block, a timestamp, and transaction data. This method of recording ensures that the blockchain is inherently secure against tampering.

The oldest blockchain has been hiding in the *New York Times* since 1995 (Czuleger 2022). Haber and Stornetta implemented their timestamping scheme by launching a service named Surety, which uses its AbsoluteProof software to timestamp digital documents. Clients generate a hash of their document, which is then sent to Surety's servers for timestamping, creating a seal. To ensure the integrity and trustworthiness of its internal records, Surety adopts a unique approach: it aggregates the hashes of all new seals added to its database weekly into a single hash value. This cumulative hash is then published in the *New York Times*, within the classified section under

**NOTICES &
LOST AND
FOUND**
(5100-5102)

Universal Registry Entries:
Zone 2 -
dS8492cgVOFAoP9kyE1XzMOrQ
HgEwzkVbVafNvlkUz99qvq8/ME
p5y9EFSG8XxzMBalGQQ==
Zone 3 -
JnFCg+HCmvhj8GmmUP7VZna71
NgZup/RfuKUQNzCHWXMuqLK
durxHQV5pSHLqBGPRly+mg==
These base64-encoded values represent the combined fingerprints of all
digital records notarized by Surety
between 2009-06-03Z 2009-06-09Z.
www.surety.com   571-748-5800

Figure 6.1 The oldest blockchain, which serves the purpose to prove the authenticity of data by the software AbsoluteProof. The hash is aggregated from data added to their database weekly.

"Notices & Lost and Found." This practice, which has been ongoing weekly since 1995, serves as a public and verifiable record, reinforcing the legitimacy of Surety's timestamping process. Despite its simplicity, this method is extremely difficult to compromise. In order to undermine the credibility of AbsoluteProof, a malicious actor would need to print a large number of *New York Times* editions containing a falsified hash and ensure these editions are distributed widely (Figure 6.1).

The concept of blockchain was first proposed in 1993, utilizing Merkle Trees to efficiently consolidate data into a single block (Bayer et al. 1993). In 2008, Bitcoin was introduced (Bitcoin 2008), and the first Bitcoin was issued in 2009, marking the birth of the first decentralized digital currency. Although decentralized systems are also distributed systems, they do not have specific decision-making nodes. Instead, all system decisions are made through consensus among all nodes.

Reaching consensus in such systems is challenging. A well-known issue related to this is the Byzantine Generals Problem (Lamport et al. 2019), which describes a scenario where all generals attacking Byzantium must receive and act upon the same information, without allowing traitors or spies to transmit false intelligence. Any miscommunication could lead to failure in their siege. A potential solution to this is practical Byzantine Fault Tolerance (pBFT)(Castro and Liskov 1999), which operates in four stages: (1) a user sends a request to a node; (2) the node broadcasts the message to all nodes; (3) all nodes that receive the message must reply to the initial node; (4) the initial node verifies all responses, ensuring that at least m nodes (the maximum number of faulty nodes allowed) have received the correct message before recording it (Figure 6.2).

However, this method is highly susceptible to Sybil attacks (Douceur 2002), where a malicious user creates multiple false identities to disrupt the consensus process. Thus, there's a pressing need for a method to verify the legitimacy of identities. Proof

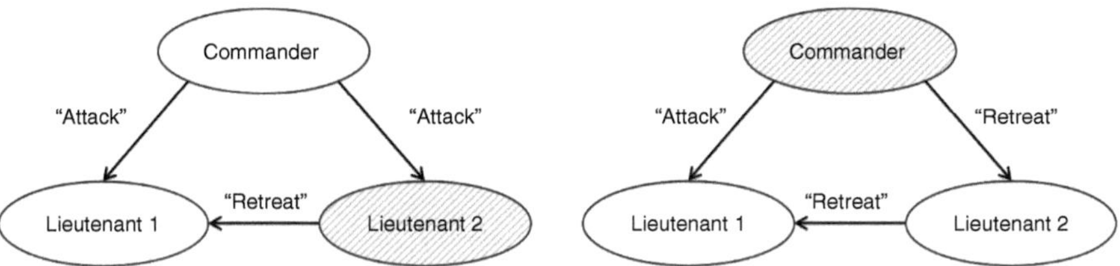

Figure 6.2 The Byzantine Generals Problem. All generals must receive correct commands so as to make a successful siege.

of Work (PoW) was proposed in 1993 as a solution (Dwork and Naor 1992), requiring all message transmitters to earn their place in the node network through work. This shifts the voting criteria in the consensus process from the number of accounts to computational power, making Sybil attacks expensive and ineffective.

Bitcoin (2008) requires network participants, called miners, to solve a complex mathematical puzzle, a process that requires significant computational effort and energy consumption, so as to prevent double-spending and ensure the security and integrity of the blockchain.

When a miner solves the puzzle, they are allowed to add a new block of transactions to the blockchain. This block contains a special transaction that rewards the miner with newly minted bitcoins (the block reward) and the transaction fees from the transactions included in the block. The difficulty of the puzzle adjusts dynamically, ensuring that the average time between new blocks remains approximately 10 minutes, regardless of the total computational power of the network.

The PoW algorithm used in Bitcoin is based on the SHA-256 hash function. Miners generate numerous hash values by altering a nonce (a number that is used only once) in the block header until they find a hash that meets the network's current difficulty target. This target ensures that the hash of the block's header begins with a certain number of zero bits. The first miner to find a valid hash announces the new block to the network, which then verifies the validity of the block and adds it to their version of the blockchain (Figure 6.3).

There have recently been other forms of PoWs. For example, in hard-drive-based PoW systems, participants, often called farmers rather than miners, allocate a portion of their disk space to support the network. The process involves storing a large number of possible cryptographic solutions on the hard drive. When the network needs to reach consensus, it challenges farmers to provide a solution to a specific problem, which they can quickly retrieve from their pre-stored solutions if they have the correct one.

One of the most well-known implementations of hard-drive-based PoW is Chia (Network 2018), which uses a concept called Proof of Space and Time. Proof of Space requires users to allocate unused disk space to store a collection of cryptographic numbers, known as plots. When the network broadcasts a challenge for the next block, farmers scan their plots to find a hash value that is closest to the challenge. Proof of Time, on the other hand, ensures that there is a minimum time gap between blocks, adding security against various attacks and manipulation.

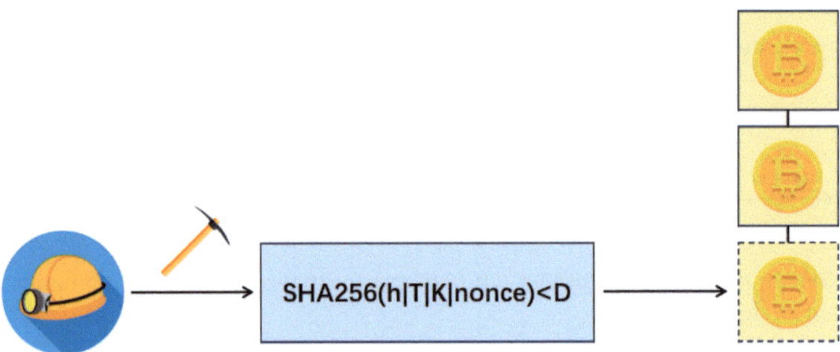

Figure 6.3 In PoW consensus mechanisms, miners are required to solve a cryptographic hash puzzle in order to append new blocks to the chain. This ensures the integrity and security of the blockchain, mitigating against malicious tampering attempts. The hash puzzle depicted in the diagram corresponds to that utilized in the Bitcoin network. Here, "h" represents the hash of the latest block in the longest chain, "T" denotes the transactions to be recorded, "K" signifies the public key, "D" represents the prevailing difficulty level, and "nonce" refers to the solution sought by the miner.

Ethereum (Wood 2014), launched in 2015 by Vitalik Buterin and his team, significantly expanded the capabilities of blockchain technology beyond its initial function as a digital currency ledger, as exemplified by Bitcoin. Ethereum's introduction of smart contracts, self-executing contracts with the terms of the agreement directly written into lines of code, has been a game-changer. These contracts automatically enforce and execute the terms of an agreement based on predefined rules, facilitating a wide range of decentralized applications (dApps) without the need for intermediaries.

The Ethereum blockchain hosts these smart contracts, allowing developers to create and deploy applications that can perform a variety of functions, from creating decentralized financial instruments to issuing and managing digital assets. The execution of these contracts and applications is managed by the Ethereum Virtual Machine (EVM), a powerful, decentralized computing platform that ensures the integrity and security of data without any risk of downtime, censorship, or third-party interference.

Despite Ethereum's innovative advancements, the platform has encountered scalability challenges, high transaction fees, and network congestion, primarily due to its reliance on the PoW consensus mechanism. To address these challenges, the Ethereum community is in the process of upgrading to Ethereum 2.0, which includes a transition to Proof of Stake (PoS). This upgrade aims to enhance network efficiency, reduce transaction costs, and significantly decrease the blockchain's energy consumption, making it more sustainable and scalable.

Ethereum's impact extends beyond technical innovations; it has catalyzed the development of an entirely new sector within the tech industry known as decentralized finance (DeFi), alongside fostering new forms of digital governance and collective organization through decentralized autonomous organizations (DAOs).

Within the contemporary landscape of blockchain technology, its application has transcended the initial confines of digital currencies and ledgers, venturing

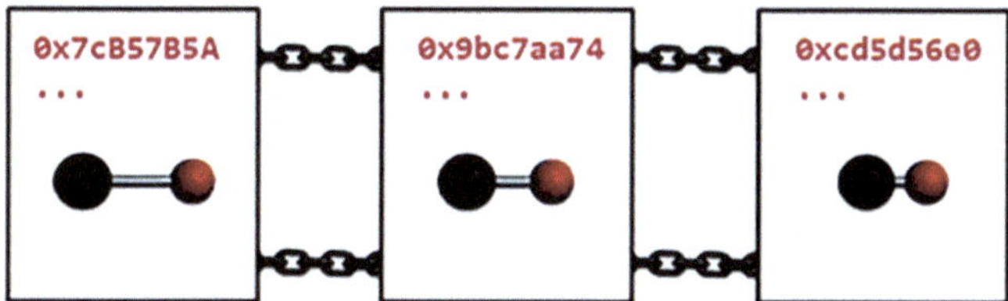

Figure 6.4 Material scientists have adopted blockchain in molecular dynamics.

into the realm of computational chemistry. A recent innovation (Hanson-Heine and Ashmore 2020, 2023) has seen scientists successfully conducting molecular dynamics simulations directly on the Ethereum blockchain's virtual machine. By employing a harmonic potential model to simulate the vibrations of carbon monoxide molecules, this advancement not only showcases the versatility of blockchain technology beyond its original financial applications but also unveils the significant potential of blockchain as a computational platform. This breakthrough paves new pathways for enhancing the transparency and accessibility of computational science (Figure 6.4).

Moreover, the application of this technology holds the potential to span across multiple disciplines, including physics, chemistry, and biology, thereby providing a novel, decentralized platform for computational simulations. This implies that through the utilization of blockchain technology, scientific computational results can be shared and verified on a global scale, significantly augmenting the reproducibility and transparency of scientific research findings. In the face of the ongoing reproducibility crisis confronting the scientific community, this represents a potential solution.

By elucidating the capacity of blockchain technology to facilitate complex computations within the field of computational chemistry, this innovation not only underscores the multifaceted utility and adaptability of blockchain but also heralds its potential in driving forward scientific research. The global sharing and verification of scientific computations via blockchain technology offer a promising avenue toward addressing the challenges of reproducibility and transparency in research, marking a significant stride toward the advancement of scientific inquiry and innovation.

6.2 Laboratory Chemical Management and Safety

In the realm of automated chemical processes, stringent management of laboratory chemicals is a critical factor for maintaining safety and efficiency. Given the precise manipulation of a plethora of chemicals and reagents in automated processes, a lack of a robust management system may lead to significant incidents, such as the loss or misuse of hazardous substances. This not only jeopardizes the safety of laboratory personnel but also poses a potential irreversible threat to the environment. Consequently, establishing and sustaining a comprehensive chemical management and tracking system is paramount in preventing accidents, ensuring the smooth progression of experiments, and safeguarding both laboratory personnel and the environment. This involves, but is not limited to, rigorous oversight of chemical procurement, storage, usage, and disposal, as well as the implementation of effective emergency measures to address potential chemical spills or other related emergencies.

The paper "Towards Blockchain-Driven, Secure and Transparent Audit Logs" (Ahmad et al. 2018) introduces BlockAudit, a novel system that integrates audit logs with blockchain technology to enhance the security and integrity of audit trails in Online Transaction Processing (OLTP) systems. The motivation behind BlockAudit stems from the vulnerabilities present in traditional audit logs, which are susceptible to various attacks that can compromise the integrity of OLTP systems. These attacks include exploiting weaknesses in the logging site, compromising database security, and manipulating audit logs to cover up unauthorized activities.

To address these challenges, the authors propose BlockAudit as an end-to-end solution that raises the security standards of OLTP systems while ensuring operational consistency. The key objective of BlockAudit is to replicate the state of audit logs across multiple entities related to the OLTP application, maintaining an append-only ledger that is robust against internal and external attacks on the database. By distributing audit logs across multiple peers, BlockAudit aims to enable the detection of unauthorized activities and restore the system to a stable state in case of security breaches.

The implementation of BlockAudit involves leveraging existing audit log generation systems, such as those provided by ClearVillage Inc., and integrating blockchain technology to store the audit log securely. The system architecture includes web and mobile clients, a business logic layer, an object-relational mapping (ORM), and a database. By moving the searchable audit log to the blockchain, applications can continue to perform database queries with minimal modifications while benefiting from the security features of blockchain technology.

One of the critical components of BlockAudit is the consensus protocol employed among peers to agree on the sequence of transactions and the state of the blockchain. Various consensus algorithms, such as PoW, proof of stake (PoS), and Byzantine fault tolerance (BFT), can be utilized to ensure the integrity and immutability of the audit log stored on the blockchain.

Experimental evaluations of BlockAudit demonstrate its performance in terms of latency, network size, and payload size. The results show that as long as the network size remains within a certain threshold, the latency remains negligible even with varying payload sizes. This indicates the scalability and efficiency of BlockAudit in maintaining secure and transparent audit logs in OLTP systems.

BlockAudit presents a promising approach to enhancing the security and transparency of audit logs in OLTP systems by leveraging blockchain technology. By combining the strengths of audit logs and blockchains, BlockAudit offers a robust solution to counter audit log attacks and ensure the integrity of transaction records in online systems.

The paper "BCALS: Blockchain-based secure log management system for cloud computing" (Ali et al. 2022) authored by Ali, Khan, Ahmed, and Jeon explores the crucial need for securing audit logs within cloud computing environments. It introduces the Blockchain-based Cloud Audit Logging System (BCALS) as a novel solution to address vulnerabilities and uphold the integrity of user activities in cloud systems.

The authors commence by providing an insightful overview of existing audit log management techniques, highlighting their strengths and limitations. Through a comparative analysis of security properties across different schemes, the paper underscores the necessity for robust security features in logging systems. Vulnerabilities in log generation, collection, communication, storage, and analysis phases are discussed, emphasizing the susceptibility of known secure logging schemes to various attacks and the importance of mitigating insider threats.

BCALS is presented as an innovative approach that harnesses blockchain technology to enhance the security and trustworthiness of audit logs in cloud computing environments. The system aims to ensure confidentiality, integrity, availability, non-repudiation, and privacy – essential security requirements for a reliable logging scheme. By utilizing blockchain for secure and tamper-proof log data storage, BCALS offers a resilient solution to combat potential threats and maintain the credibility of audit trails.

The paper also delves into the performance evaluation of BCALS, showcasing the efficiency of document publishing to Elasticsearch. Performance metrics, including time distribution for parsing, JSON creation, blockchain storage, and Elasticsearch storage, are illustrated through figures, highlighting the system's scalability and effectiveness in managing log entries under varying workloads.

Furthermore, stress testing of BCALS is discussed, focusing on log entry parsing and semantic enrichment processes. Results from stress tests reveal the computational time required for initial processing, demonstrating the system's capability to handle continuous loads and sustain performance efficiency.

The authors emphasize the critical role of secure log management in cloud computing environments and introduces BCALS as an advanced solution to address security challenges associated with audit logging. By leveraging blockchain technology and prioritizing key security requirements, BCALS offers a robust framework for ensuring the integrity and trustworthiness of audit logs in cloud computing systems.

6.3 The Problem of Data Integrity and Counterfeiting in Scientific Research

Over the past decade, experimental disciplines have experienced rapid development, particularly in the fields of biology, chemistry, and materials science, thanks to advancements in automation and related technologies. Concurrent with this swift progress, the surge in academic publications has also led to an increased incidence of practices that do not adhere to scholarly standards.

By the close of 2023, *Nature* released statistical data pertaining to the retraction of scholarly articles for the same year (Van Noorden 2023). A staggering number of over 1000 academic publications were retracted in 2023 alone, representing a nearly

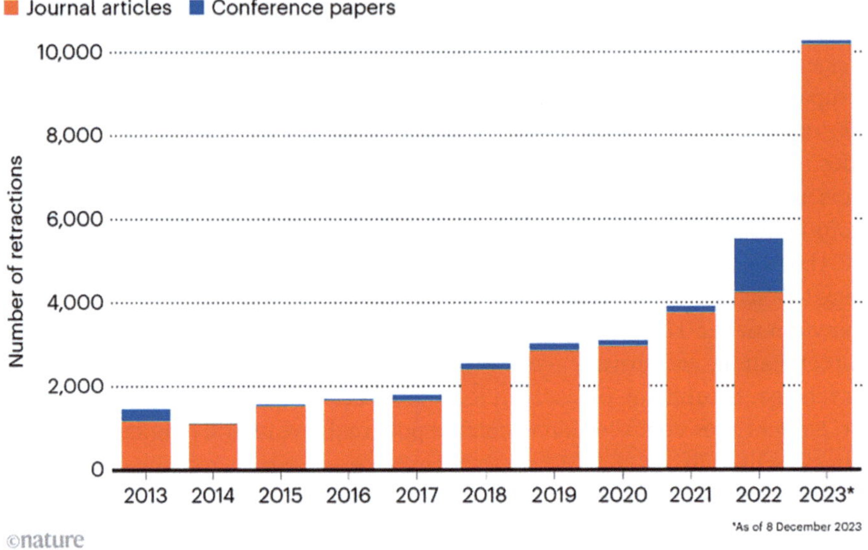

Figure 6.5 A bumper year for retractions. Source: Retrieved from *Nature* (Van Noorden 2023).

twofold increase compared to the preceding year of 2022. Notably, the majority of the retractions were attributed to a singular journal, Hindawi. In the course of the current year, Hindawi journals have retracted upward of 8,000 articles. These retractions were justified on the grounds of concerns surrounding the compromise of the peer-review process and the systematic manipulation of both publication and peer-review procedures. The investigations leading to these retractions were initiated internally by editors and further fueled by research-integrity investigators who raised alarms regarding inconsistencies in the textual content and the inclusion of irrelevant references across thousands of papers (Figure 6.5).

In a recent discovery within the realm of academic research, an unconventional paper mill has come to light, presenting a peculiar scenario (Else 2022b) Surprisingly, an extensive body of scholarly work, amounting to approximately 1,000 papers, revolves around an imaginary intersection between advanced chemistry and medical applications. Notably, these papers, albeit not all housed in readily accessible journals, assert the utilization of metal–organic framework (MOF) compounds for purposes such as cancer cell eradication and inflammation inhibition. While MOFs do possess remarkable physical properties, the notion of their medical applications appears highly implausible. The origin of suspicions arose during a perusal of PubPeer, a post-publication peer-review site. Credit is due to Sylvain Bernès, a Mexican crystallographer, who initially raised questions about some of these papers. The discovery gained momentum as recurring patterns emerged, leading to the identification of additional examples within the same journals. What heightened concerns were the utilization of bogus reference sections, where papers recycled references unrelated to their citations. This discovery led to a productive exploration of the paper mill's extensive network, wherein citation exchanges occurred through a central broker, allowing researchers to cite papers without direct knowledge of the individuals involved. This phenomenon bears resemblance to a previous *Nature* interview with an anonymous academic whistleblower named Clyde, who uncovered similar academic misconduct by tracing through shared references and networks within the same journal. These revelations underscore the existence of paper mills and the anonymous exchange of references as a means of reciprocal citation enhancement within certain academic circles.

Caution is warranted not only with obscure groups and individuals lacking academic creativity but also among Nobel laureates, as instances of data fabrication, irreproducible experiments, and article retractions have been observed. This phenomenon complicates the research landscape, heightening the difficulty of literature reviews for researchers. Even when faced with authoritative publications, scholars cannot unequivocally trust their content, necessitating the verification of authenticity. The occurrence of these issues underscores the growing challenges researchers face in navigating the integrity of scholarly work. Below are some recent examples in which the papers are found with faked data or images.

A study led by Nobel Prize–winning neuroscientist Thomas Südhof in 2023 has raised significant concerns about the validity of reported data, prompting the publisher to issue an editorial expression of concern (Lin et al. 2023). The paper, titled "Neurexin-2 restricts synapse numbers and restrains the presynaptic release probability by an alternative splicing-dependent mechanism," published in the *Proceedings of the National Academy of Sciences* (PNAS), has been flagged following a reader's concerns and subsequent investigations into potential data issues. This marks the third instance this year where Südhof's research has faced scrutiny or corrections by journals. The specific concerns revolve around numerical discrepancies in supplementary spreadsheets, with anonymous commenters on PubPeer pointing out repeated values. Despite the first author, Pei-Yi Lin, responding to critiques by providing updated spreadsheets, doubts linger, with physicist Maarten van Kampen suggesting possible fabrication due to statistically improbable identical decimals. Südhof acknowledged major inconsistencies and submitted an erratum to PNAS in an effort to rectify the record. This editorial expression of concern adds to a growing trend of high-profile retractions and corrections within the scientific community, raising questions about the reliability of research outcomes even from renowned scientists.

During Nobel Prize week of 2023, Gregg Semenza, a Nobel laureate and professor of genetic medicine at Johns Hopkins' Institute for Cell Engineering, has retracted his 10th paper due to issues with data and images (Else 2022a; Kincaid 2023). Semenza, awarded the 2019 Nobel Prize in Physiology or Medicine for discoveries related to cells' response to oxygen availability, has faced scrutiny since before his Nobel recognition. Concerns about potentially duplicated or manipulated images in Semenza's publications were raised by pseudonymous sleuth Claire Francis on PubPeer, with additional scrutiny from other observers starting in October 2020. The recent retraction in Molecular Cancer Research, titled "Procollagen Lysyl Hydroxylase 2 Is Essential for Hypoxia-Induced Breast Cancer Metastasis" (2013), marks the latest in a series of retractions. The notice cites identical images in the HIF-1α immunoblot as the reason for retraction, a claim corroborated by an internal review and acknowledged by the editors. This development adds to Semenza's previous retractions, raising questions about the integrity of his research output, and comes amid ongoing scrutiny of high-profile scientists' work within the scientific community. Semenza did not respond to requests for comment, and Johns Hopkins declined to comment on whether an investigation was underway.

Kyoto University has "punitively dismissed" researcher Kohei Yamamizu, the corresponding author of a 2017 stem cell paper, after an investigation revealed fabrication and falsification in nearly all figures except one (Yamamizu et al. 2017, 2018). The study, published in *Stem Cell Reports*, was retracted earlier this month, and the university determined that Yamamizu was solely responsible for the misconduct. Shinya Yamanaka, director of the Center for iPS Cell Research and Application (CiRA) and a Nobel laureate in stem cell biology, acknowledged a "strong responsibility" for failing to prevent the research misconduct and has received a penalty from the university. The penalty details were not specified, and Kyoto University has not responded to inquiries. The retracted paper claimed to model the blood–brain barrier using human iPSC-derived cells but lacked support due to manipulated and missing data.

Additionally, numerous articles have been retracted due to the inability to replicate their experiments.

In 2020, Frances Arnold, a Nobel laureate from Caltech, had retracted a 2019 paper on enzymatic synthesis of beta-lactams after being unable to replicate the results (Cho et al. 2019, 2020) Arnold won the 2018 Nobel Prize in Chemistry for her work on the evolution of enzymes. The retraction notice stated that efforts to reproduce the work revealed that the enzymes did not catalyze the reactions as claimed, and examination of the first author's lab notebook showed missing contemporaneous entries and raw data for key experiments. Arnold admitted fault for not paying enough attention to the submission during the Nobel Prize hoopla and apologized for the retraction.

In 2017, *Science* retracted a 2014 paper from the lab of Nobel laureate Bruce Beutler after replication attempts failed to conclusively support the original results (Berg 2017; Zeng et al. 2014). Beutler, an immunologist at the University of Texas Southwestern Medical Center and Nobel Prize winner in Physiology or Medicine (2011), reported weakened confidence in the paper's findings, prompting retraction requests from him and several co-authors. However, two co-authors disagreed, prolonging the retraction process. The journal waited for Beutler's lab to conduct another replication attempt, but the results were inconclusive. After a month-long evaluation, *Science* decided to retract the paper, titled "MAVS, cGAS, and endogenous retroviruses in T-independent B cell responses." The retraction notice cites discrepancies in the foundational observations, leading to the conclusion that the core observations and conclusions are not robust.

In 2016, research by Nobel laureate Michael Rosbash published in PLOS Biology was challenged for reproducibility by researchers at Rockefeller University, nine years post-publication. Despite subsequent attempts by the original authors to replicate the experiment, the decision was made to retract the paper (Kaushik et al. 2007, 2016). The failure in replicating the results may arise due to overly complex methods or overlooked details that the authors themselves can replicate but others cannot. This is possible in chemical synthesis. There has been publication discussing about the reactor parameters for photocatalytic reactions, in which it says that the parameters must be designed carefully. Even the shape of the reactors could alter the products of the reaction (Cañellas et al. 2024).

Also 2016, a paper in *Nature Chemistry*, co-authored by Nobel Laureate Jack W. Szostak, claimed to have found evidence supporting the idea that certain peptides could facilitate nonenzymatic RNA replication, potentially shedding light on the origins of life. However, subsequent experiments conducted by Tivoli Olsen, a member of Szostak's lab, failed to reproduce the findings. Upon review, it was discovered that the initial data had been misinterpreted, leading to false positives. The authors requested a retraction of the paper due to the errors. This marks the second retraction in six months for *Nature Chemistry*, a rare occurrence for the journal (Litovchick and Szostak 2008, 2009). In the past, Szostak had also retracted a paper from 2008 in *Proceedings of the National Academy of Sciences* due to replication issues raised by an outside researcher, Katherine Berry.

In 2015, after eight years of investigation, federal officials have found Duke University cancer researcher Anil Potti guilty of research misconduct (Potti et al. 2006, 2011). Potti's team published papers in 2006 claiming that gene expression signatures predicted a patient's response to chemotherapy, but concerns were raised, leading to Potti's suspension in 2010 and subsequent resignation. The Office of Research Integrity (ORI) concluded that Potti included false research data in a grant application, a manuscript, and nine research papers. Many of Potti's papers were retracted, and Duke faced a lawsuit. As part of a voluntary settlement, Potti neither admits nor denies the findings, and if he seeks federal funding again, his research must be supervised for five years.

Instances of irreproducible experiments can stem from various possibilities. The most egregious scenario involves the deliberate fabrication of experimental data and conclusions, where the claimed academic achievements are essentially a deception. Another possibility is that authors may intentionally or unintentionally omit seemingly inconspicuous but crucial details of the experiment. Some authors may do this to protect their interests, aiming to share academic achievements while maintaining a dominant position in the field. Alternatively, the decisive factors in the experiment may be misunderstood by the authors. For example, in the synthesis of quantum dots, the characteristics of the product may be significantly influenced by the shape of the container, a factor that might not be documented in the paper.

Furthermore, the findings of the paper might be inherently low in success rate, making replication possible but extremely challenging, thereby diminishing its scientific value. While in some cases the authors may not have had malicious intent, the difficulty in reproducing the results adds uncertainty to the development of that particular research direction and increases the difficulty of assessing the authenticity of experimental work. These issues can be broadly categorized as authenticity problems.

Another kind of problems is originality. While historically receiving less attention, it has also been at the center of notable disputes. A prominent example involves the discovery of graphene. Despite general acknowledgment attributing the discovery to two scientists in 2004, debates persist, with some reservations citing references to graphene in literature from the 1960s and 1970s (Berger et al. 2004; Geim 2010). These disputes may stem from ambiguous language in earlier papers or the disappearance of the original material over time.

6.4 Blockchain in the Autonomous Laboratory

Based on the aforementioned discussion, especially in experimental disciplines within the research domain, there is a need for a method that comprehensively records experiment timelines, protocols, and data. This approach ensures both the transparency and authenticity of data while providing a means to attribute discoveries through timestamped records.

One emerging solution in recent years is BiaeP (Xu et al. 2020), introduced by researchers at The Chinese University of Hong Kong (Shenzhen). BiaeP stands for The Blockchain Integrated Automatic Experiment Platform. This platform combines blockchain technology, cloud storage, and automated experimentation to address the goal of ensuring complete and transparent records of experimental protocols and data. The primary objective is to maximize the reproducibility of experimental protocols by leveraging the immutability and traceability features of blockchain technology, along with the efficiency of cloud storage and the automation capabilities of experimental platforms.

The BiaeP system, as depicted in Figure 6.6, is an advanced integration of two primary sub-platforms: the Ethereum Blockchain & Cloud Storage Platform and the Automatic Experiment Platform. These components are interconnected through a sophisticated information transmission and storage protocol, termed the BIP file.

The BIP file serves a dual purpose: it acts as a repository for experiment parameters within the Ethereum Blockchain & Cloud Storage Platform and conveys procedural instructions to the Automatic Experiment Platform. Comprising two main sections – BIP instructions and BIP results – the file meticulously documents the experimental procedures and configurations. BIP instructions detail the necessary steps and configurations, employing a JSON-like format that lists required instruments and chemical reagents, as well as outlining the complete experimental procedure from preparation to result analysis. This structured approach ensures that every parameter required by the Automatic Experiment Platform is accounted for. Conversely, BIP results focus on documenting the outcomes of experiments, supporting raw data input to facilitate reproduction by other researchers, irrespective of their hardware configurations.

Figure 6.6 The BiaeP system. The system comprises Ethereum, a cloud storage platform, and an automated experimentation platform. The experimentation platform utilizes experiment files in BIP format, which will be stored on a third-party cloud storage platform. The hash of these files will be uploaded to Ethereum to ensure the originality of the experiments and provide a time-stamped traceability. This process also guarantees the integrity of the experimental designs and data against tampering.

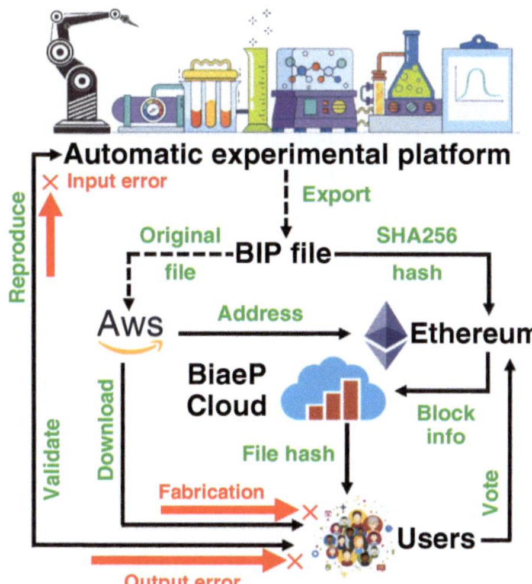

The BIP file's design emphasizes readability and ease of use, enabling researchers without access to an Automatic Experiment Platform to review and reference the experimental protocols. This approach significantly reduces the risk of reproducibility failures, as the Automatic Experiment Platform is engineered to interpret and execute BIP instructions autonomously, eliminating human error.

The development of the Automatic Experiment Platform draws inspiration from cutting-edge technologies in flow chemistry and robotic automation. Its design addresses the lack of a unified standard for laboratory instruments by introducing a flexible communication protocol. This protocol, equipped with specific sockets for a wide range of instruments, allows for the easy integration and adaptation of various laboratory apparatus. Each socket specifies the conditions under which the instrument operates, such as temperature and duration, and can be accessed through BiaeP reserved words, ensuring the platform's adaptability and ease of extension.

Constructed with these principles, the benchmark Automatic Experiment Platform is designed for maximum adaptability and usability. Positioned within a four-sided fume hood, the platform includes several permanent functional units alongside two optional units that can be customized based on the specific needs of the research. This modular design allows for the seamless integration of new instruments and technologies, underscoring the system's potential for future expansion and adaptation to evolving experimental requirements.

The inherent flexibility and advanced capabilities of the BiaeP system underscore its potential to revolutionize experimental practices. By automating the execution of complex experimental protocols, the system not only enhances the accuracy and reproducibility of research outcomes but also paves the way for new discoveries in the realm of scientific experimentation.

While there exist more advanced and faster blockchain platforms, such as EOS, their developmental status and security vulnerabilities render Ethereum the preferable choice for this application. Ethereum's requirement for real identity registration serves as a deterrent against potential malicious activities.

The Ethereum Blockchain, with its inherent unforgeability and tamper-resistance, represents an ideal solution for securing research records. However, the platform's architectural limitations pose challenges for directly writing large files, such as those generated by electron microscopy, which can be megabytes in size. This is primarily due to the prohibitive cost associated with Ethereum's gas fees, charged for each byte of data written to the blockchain. As a compromise, a cloud storage (AWS S3 storage, as shown in Figure 6.7(a)) system is utilized for the bulk storage of data, while the Ethereum Blockchain is reserved for verifying authenticity through the recording of digital fingerprints.

For instances where the disclosure of experimental details is not desired, the BIP file stored in the cloud may be encrypted using the public key associated with the authors' Ethereum address. Nonetheless, it is mandatory to upload the SHA256 hash value of the BIP file to the blockchain to ensure the integrity and authenticity of the record.

Upon completion of a successful experiment in the experimental stage, as shown in Figure 6.7(b), the BiaeP system facilitates the automatic generation of a BIP file. Authors are then presented with the option to either keep the experiment details private or make them publicly accessible. In the case of non-disclosure, the BIP file is encrypted and stored, while its SHA256 hash value and a direct URL to the storage location are recorded on the Ethereum Blockchain. This process ensures the permanent and verifiable storage of the record. Once published, the BIP file becomes accessible to the research community, enabling the verification of authenticity and fostering an environment conducive to reproducibility and academic integrity.

The system also incorporates a voting and commentary mechanism, allowing researchers to share their findings on the reproducibility of experiments. This peer-review-like function is bolstered by an identity verification requirement for all users, ensuring the credibility of contributions. The option to publish the BIP file at a later date remains open to authors, provided they can decrypt the file to verify its consistency and ownership. This approach not only facilitates the reproducibility of research but also promotes a collaborative and transparent scientific community.

The Blockchain Information Protocol for Experiments (BiaeP) system is designed with robust security measures to safeguard against potential threats to the integrity of scientific research records. This discussion outlines the anticipated vulnerabilities of BiaeP and the preventive strategies embedded within its architecture.

In scenarios where users opt to publicly disclose a BIP file from the onset, the system employs an innovative approach to ensure the authenticity and originality of the data. Contrary to conventional methods that rely on confidentiality to protect these attributes, BiaeP adopts an open-access model, underpinned by cryptographic techniques, enabling universal supervision and guaranteed authenticity.

Three principal attack vectors have been identified, as illustrated in Figure 6.7c. The first concerns unauthorized alterations to the original file stored in the cloud. Due to the encryption algorithm's avalanche effect, any minor modification to the file induces a significant discrepancy in its hash value, consequently failing the checksum validation process. The second and third vectors pertain to the fabrication

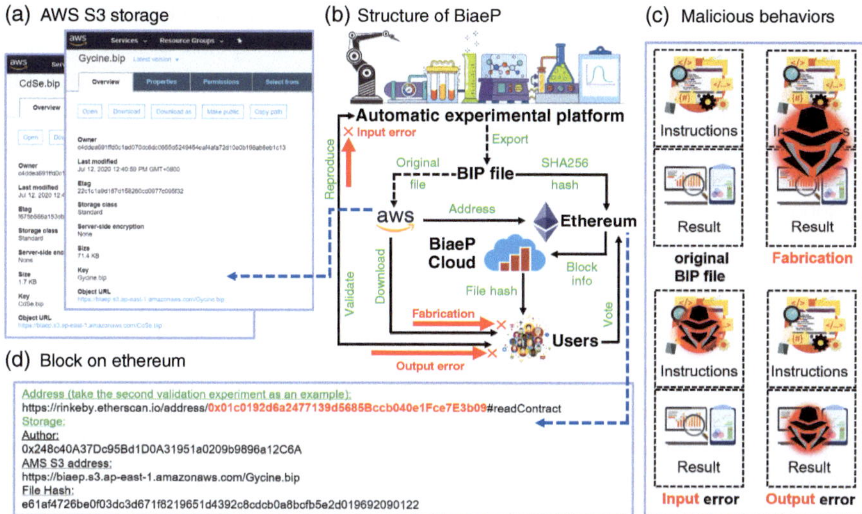

Figure 6.7 The working methodology of BiaeP. (a) Storage of demo files on AWS S3. (b) The SHA256 hash of the BIP file is recorded in a smart contract on the Ethereum network (BiaeP) if needed, along with its AWS S3 address. Users have the choice to upload the original file or encrypt it asymmetrically using the public key of their Ethereum address. Subsequently, other users can access the file and verify its authenticity through the automated platform. (c) Various types of malicious behavior can be identified by researchers, including discrepancies such as unavailable experimental methods, hash value mismatches (indicating modified BIP files in AWS), or alterations in experimental data (resulting in modified results). (d) Information on blockchain blocks. Details regarding the block on the Rinkeby test net, where the demo file "Glycine.bip" was written, are provided.

of experimental instructions and results prior to cloud upload. These fraudulent actions are expected to be exposed through discrepancies in reproduction results by other researchers.

It is important to acknowledge that BiaeP cannot autonomously detect deliberately falsified BIP files. The system thus relies on the collaborative efforts of the research community and the capabilities of the automatic experiment platform. Feedback mechanisms, facilitated by voting and commenting functions in the Ethereum Blockchain smart contract, serve as a means for collective scrutiny and validation, enhancing the credibility of records with increased user engagement and subsequent reproductions.

Furthermore, the system provides a robust defense against plagiarism accusations. In such instances, the blockchain timestamp of the BIP hash unequivocally establishes the chronology and provenance of the contested research, offering protective evidence for the implicated party.

Even if the initial decision was to withhold the BIP file from public view, the aforementioned protective mechanisms remain effective upon the file's decryption and controversy emergence. The on-chain hash of the BIP file serves as a timestamped PoW, nullifying fraud accusations and facilitating the clear identification of the experiment's original innovator.

Financially, the cost of engaging with the BiaeP system is relatively modest. With the average price of Ethereum at approximately \$300, the expense associated with creating a record on the system is about \$2.4. This includes the fees for smart contract execution and the linkage of the hash and storage URL to the Ethereum mainnet. Additionally, users benefit from free cloud storage services provided by platforms such as Google Cloud and Amazon S3, under their free usage tiers, capable of hosting over 5000 BIP files.

The requirements for implementing a BiaeP environment vary significantly, dependent on the specifics of the automatic experiment platform. While basic service access without an automatic experiment platform is complimentary, the development and integration of such platforms can be intricate and costly, necessitating the replacement of incompatible instruments and the creation of bespoke drivers for BIP socket connectivity.

Despite these challenges, the advantages of adopting BiaeP are manifold. It offers irrefutable evidence of experimental authenticity, facilitates global engagement with the latest scientific endeavors, and enhances the efficiency and accuracy of experimental workflows through laboratory automation, all at a competitive cost. These attributes position BiaeP as an attractive proposition for the scientific community, promising to advance the integrity and efficiency of research practices.

A time-dependent precursor injection strategy in the synthesis of CdSe quantum dots was used as an example to illustrate this protocol. The methodology behind this experiment leverages optimization for achieving a more uniform size distribution of quantum dots, building upon previously reported optimized injection techniques. Through numerical simulations, better injection strategies are explored, which are then executed and evaluated using the BiaeP system. The outcomes of these simulations, represented by solid lines in the referenced figure, serve as benchmarks for subsequent experimental repetitions.

The apparatus for conducting quantum dot synthesis via BiaeP is detailed in Figure 6.8, where the directionality of solution flow through the system is indicated by arrows on the pipes. Given the propensity of precursor solutions to degrade over time, their onsite preparation is necessitated, facilitated by robotic arms and a solid feeding module. The hot injection method, akin to the process described by Lai et al. (2018) for zinc blende seed synthesis, is employed for the preparation of CdSe quantum dots. This process has been automated and executed twice to underscore the system's capability in ensuring the authenticity of experimental outcomes.

A key component of the BiaeP system's infrastructure is the deployment of an Ethereum-based smart contract on the Rinkeby test chain. This smart contract facilitates the immutable recording of the BIP file on the blockchain, safeguarding against unauthorized modifications. Access to the hash code of the BIP file, the AWS address where the BIP file is stored, and the solidity code of the smart contract is provided, enabling transparency and verifiability.

The integral function of the BiaeP system is predicated on the automatic experiment platform's ability to faithfully reproduce experiments from identical BIP files. This ensures the reliability and repeatability of experimental processes, which are fundamental to the system's objective of enhancing the credibility and authenticity

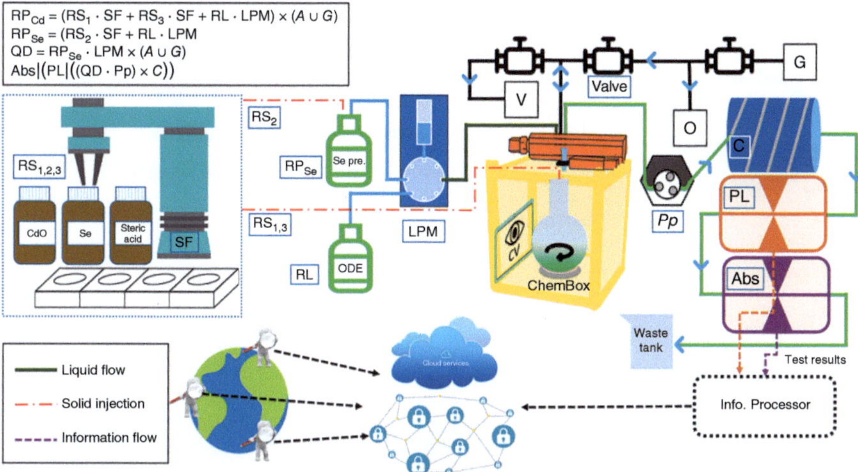

Figure 6.8 The symbolic BIP instruction, denoted by an abbreviation in the top-left corner, outlines the reaction procedure for the linear injection method employed in the synthesis of quantum dots. The various modules involved in the process include V (vacuum pump), O (gas treatment), G (gas supplier), LPM (liquid preparation module), Pp (peristaltic pump), RS (solid reagents), RL (liquid reagents), SF (solid feeding module), C (cooling module), PL (PL measurement), and Abs (absorption measurement). Subsequently, all gathered information undergoes automated processing on BiaeP and, upon confirmation by the authors, will be made available worldwide.

of scientific research. Through the implementation of such a system, the study demonstrates a novel approach to securing and validating experimental protocols in the domain of scientific research.

6.5 Symbolic Representation of Experiments

We propose taking this further by introducing a 3D visualization approach to display experimental processes more dynamically, called the CubeRoot. As illustrated in Figure 6.9a, we present a typical example of experiment steps constructed across three dimensions – time, cost, and data amount – and distinguished by two types of colors. Time costs encompass both reagent and power consumption costs, while data costs include liquid information, parameter data for each procedure step, and characterization data from the steps. Type I colors denote the reactor shell, whereas Type II colors fill actions such as operations, characterizations, and additions of specific reagents. Essential for optimizing experiment parameters is the in situ experiment information captured through a new definition termed "time-capture" (TC). TC chronologically records diverse platform data and multimodal in situ characterizations to construct high-fidelity digital twins of complex systems. It enables in situ documentation of experimental conditions, reactions, parameters, progress, and raw characterization data with multimodal sensors tracking physicochemical transformations in real-time at high fidelity across all vessels.

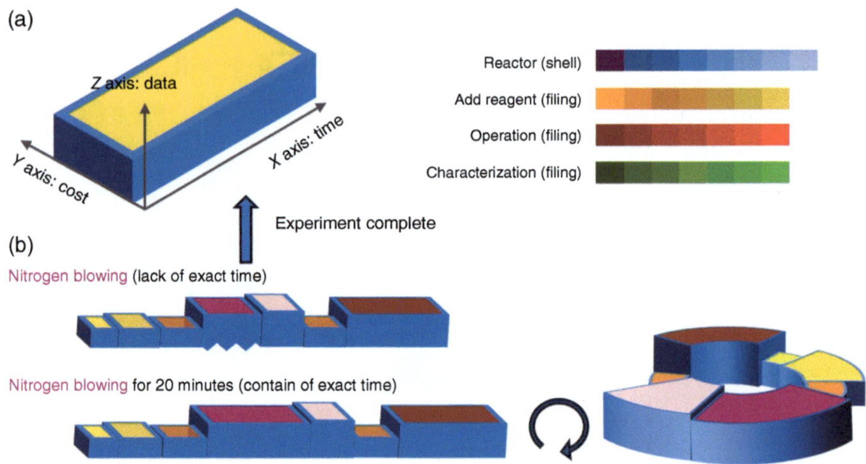

Figure 6.9 The basis of the CubeRoot. (a) Introduction of the parameters in a standard experiment step and the recorded information during experiments in time-capture. (b) An example of experiment is a combination of steps, and with or without complete experiment time, and the correlation between experiment steps and root steps.

Combining these basic blocks as demonstrated in Figure 6.9b, we visualize steps within a single experiment. For experiments lacking specific timing details for each step, we can default to using models of initial length for visualization purposes. For experiments with known details, model sizes can be adjusted based on duration – with longer times resulting in longer models. These bar-shaped models can be coiled into a ring for compact display within limited spaces; we suggest fixing a time period (e.g. 2 hours) per circle to approximate the duration of each step efficiently.

Figure 6.10 is an example of visualizing experimental data using CubeRoot. This experiment consists of two parts. In the first part, different chemicals were added to six bottles in varying proportions, and in the second part, the Raman spectra of these bottles were measured individually. As we can see in the figure, the CubeRoot of each step is arranged clockwise from this perspective, and the thickness representing the measurement steps is significantly greater than that of the pipetting steps. This is because the characterization measurement steps generate a larger amount of data

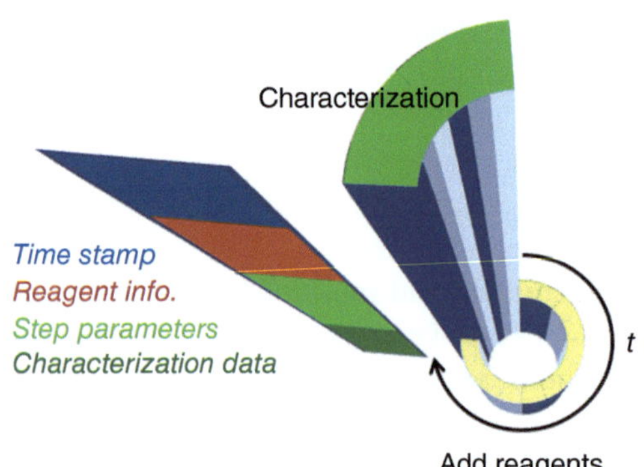

Figure 6.10 An example for the CubeRoot for a template experiment. The experiment contains two parts. First, the reagents are added to separate bottles. Then, the Raman spectra of bottles of different reagents ratios are taken. Since the Characterization steps contain more data, they are much thicker than the Add Reagent steps.

per unit of time. This can be observed in the time slices, where the characterization measurement steps include a significant amount of additional spectral data, whereas the pipetting steps include only the type and volume of liquid being moved, which is much smaller in data volume compared to the spectra.

Like BIP, CubeRoot also has a custom experiment definition method and data storage format. We define this experimental protocol file as an FF file, and the readable format of this type of file follows the writing specifications of the JSON format. Different steps are distinguished by different "functionCode" fields, and each step has its corresponding sub-fields. For example, in the case of pipetting, the fields are as follows:

```
{"functionCode": "AspireLiquid",
 "blockType": str,
 "num": int,
 "volume": float},
```

Similar to FF, the data file of the CubeRoot is also of JSON format. For example, the Characterization step is saved in the following format:

```
{"materialsList": [{"name": str, "amount": float}, ...],
 "spectrumInfo": {
   "spectrumX": list[float],
   "spectrumY": list[float]},
"ramanParameters": {...},
...},
```

The CubeRoot not only offers an intuitive 3D representation of experimental steps and data, but its storage structure also facilitates the subsequent replication of experiments. In the structure of the right part of Figure 6.9b, slicing along the radial direction of the CubeRoot reveals the experiment's information at that specific time point, including real-time control parameters, equipment readings, and characterization data. This essentially creates a complete digital twin of the experimental process. Compared to BIP files, this format enhances the concept of temporal slices in experiments. Similar to the BIP files in Biaep, this data storage format can also be asymmetrically encrypted and stored on the blockchain, allowing future generations to verify, replicate, and optimize the experiments.

6.6 Challenges and Limitations

6.6.1 Standard Compilation for Experiment Methods

Integrating blockchain technology into automated experimentation undeniably offers a robust safeguard for the security and credibility of data. To achieve this objective, it is imperative to standardize and meticulously compile experimental protocols, ensuring uniformity and precision in descriptive standards. This effort not only facilitates the reproducibility of experiments but also simplifies the process for other researchers to read and comprehend the methodologies involved. Despite the significant advancements achieved by Large Language Models (LLMs) and other AI technologies, there remains a substantial scope for development in their integration with blockchain technology. LLMs excel in processing and analyzing vast datasets, whereas blockchain technology specializes in securing data integrity and trustworthiness. An effective amalgamation of these two domains could significantly advance automated experimental processes, thereby enhancing the reliability of experimental outcomes and the security of data.

Looking forward, it is reasonable to anticipate an increase in researchers exploring this interdisciplinary field, aiming to achieve a seamless integration of blockchain technology and LLMs. This exploration could propel automated experimentation to new heights, not only increasing the efficiency of experimental processes but also bolstering the credibility and impact of the scientific research domain at large. The union of blockchain technology and LLMs heralds a promising future for the elevation of experimental standards, offering a pathway to more secure, reliable, and transparent scientific inquiry.

6.6.2 High Cost for PoW and PoS

When incorporating blockchain technology into automated experimentation, we are confronted with the necessity for consensus mechanisms such as PoW and Proof of Stake (PoS). These mechanisms provide a solid foundation for the security and trustworthiness of blockchain but also introduce higher costs. The PoW mechanism requires participants to solve computational puzzles to verify transactions and maintain network security. However, this approach demands substantial computational resources and energy consumption, leading to significant costs. For automated experiments, the adoption of the PoW mechanism could restrict the scale and frequency of experiments due to its high operational costs. In contrast, the PoS mechanism reduces costs by allowing holders to participate in network verification by staking a certain amount of tokens. Nevertheless, PoS also poses a risk of centralization, as nodes with a large number of tokens might wield excessive influence over the network. In summary, the introduction of blockchain technology into automated experimentation necessitates a careful balance between cost and security. Continuous exploration and innovation in consensus mechanisms are crucial to advance the efficiency and reliability of experiments.

6.6.2.1 Data Storage Safety

Traditionally, due to the often large size of experimental data and in pursuit of ensuring data security and credibility, the choice to store data on third-party platforms is prevalent. However, this method harbors a potential risk: should the third-party platform's data be tampered with, the experimental data stored therein also faces the risk of alteration. Moreover, once data is altered, it is generally only possible to disprove its authenticity rather than restore it to its original state. To counter this challenge, additional measures are necessitated.

Firstly, data backup is crucial. Through regular backups, we can ensure that data can be quickly restored to a previous state in the event of tampering or loss. To enhance the reliability and security of backups, it is advisable to store backup data in multiple locations to prevent single points of failure.

Secondly, the adoption of a distributed storage solution, which stores data across multiple nodes rather than relying on a single third-party platform, should be considered. In this way, even if data on some nodes is compromised, the integrity and credibility of data on other nodes can still be maintained. Furthermore, the decentralized nature of blockchain technology can be leveraged to construct a decentralized data storage network. In this network, each node can participate in the storage and verification of data, thereby ensuring its security and credibility.

In summary, to safeguard the security and credibility of data in automated experiments, a series of measures must be implemented to mitigate the risk of data tampering on third-party platforms. Through strategies such as data backup, distributed storage, and decentralized storage, the reliability and security of data can be enhanced, providing a more robust guarantee for automated experimentation.

6.7 Conclusion

In the journey toward revolutionizing the autonomous laboratory environment, the integration of blockchain technology emerges as a pivotal innovation, promising to redefine the paradigms of data integrity, security, and collaboration in scientific research. Chapter 6 has meticulously explored the multifaceted role of blockchain in ushering in a new era of experimental synthesis, marked by unparalleled efficiency, transparency, and trustworthiness.

The exploration began with a deep dive into the core principles of blockchain technology, elucidating its potential to serve as an immutable, decentralized ledger that can securely record and store experimental data. This technological foundation paves the way for addressing the pressing challenges of chemical management and safety in laboratories, ensuring that every material and its handling are accurately documented and verified, thereby minimizing risks and enhancing research outcomes.

Central to our discourse was the pervasive issue of data integrity and counterfeiting in scientific research – a dilemma that threatens the very fabric of scientific discovery and innovation. By leveraging blockchain's inherent features, we outlined a visionary framework for safeguarding experimental data against falsification, ensuring that every piece of data remains untainted and verifiable across its lifecycle.

The application of blockchain within the autonomous laboratory setting was depicted as not just a theoretical possibility but as an imminent reality. By interweaving blockchain with AI and robotics, we envisioned a laboratory environment where data authenticity and safety are not mere aspirations but tangible realities, fostering a culture of trust and collaboration among researchers.

However, the path to realizing this vision is not devoid of challenges and limitations. From technical and scalability hurdles to regulatory and adoption barriers, the journey is fraught with complexities that demand innovative solutions, interdisciplinary collaboration, and a steadfast commitment to research integrity.

As we look toward the future, the integration of blockchain technology in autonomous laboratories holds the promise of transforming scientific research. It not only offers a robust defense against the scourge of data counterfeiting but also heralds a new age of transparency, efficiency, and safety in the scientific domain. The success of this endeavor will require not just technological prowess but also a collective will to reimagine the foundations of scientific inquiry and experimentation.

In conclusion, while challenges abound, the potential rewards of a blockchain-powered anti-counterfeiting experimental data system are immense. As we venture forward, it is imperative that the scientific community, technologists, and policymakers join forces to navigate these challenges, unlocking new horizons of knowledge and innovation. Through this collaborative effort, we can ensure that the autonomous laboratories of the future are not only more efficient and autonomous but also bastions of data integrity and trust.

References

Ahmad, A., Saad, M., Bassiouni, M., and Mohaisen, A. (2018). *Towards blockchain-driven, secure and transparent audit logs.* Paper presented at the Proceedings of the 15th EAI International Conference on Mobile and Ubiquitous Systems: Computing, Networking and Services.

Ali, A., Khan, A., Ahmed, M., and Jeon, G. (2022). BCALS: blockchain-based secure log management system for cloud computing. *Transactions on Emerging Telecommunications Technologies* 33 (4): e4272.

Bayer, D., Haber, S., and Stornetta, W.S. (1993). *Improving the efficiency and reliability of digital time-stamping.* Paper presented at the Sequences II: Methods in Communication, Security, and Computer Science.

Berg, J. (2017). Editorial retraction. *Science* 358 (6362): 458–458.

Berger, C., Song, Z., Li, T. et al. (2004). Ultrathin epitaxial graphite: 2D electron gas properties and a route toward graphene-based nanoelectronics. *The Journal of Physical Chemistry. B* 108 (52): 19912–19916. https://doi.org/10.1021/jp040650f.

Cañellas, S., Nuño, M., and Speckmeier, E. (2024). Improving reproducibility of photocatalytic reactions—how to facilitate broad application of new methods. *Nature Communications* 15 (1): 307.

Castro, M. and Liskov, B. (1999). *Practical byzantine fault tolerance.* Paper presented at the OsDI.

Cho, I., Jia, Z.-J., and Arnold, F.H. (2019). Site-selective enzymatic C–H amidation for synthesis of diverse lactams. *Science* 364 (6440): 575–578. https://doi.org/10.1126/science.aaw9068.

Cho, I., Jia, Z.-J., and Arnold, F.H. (2020). Retraction. *Science* eaba6100. https://doi.org/10.1126/science.aba6100.

Czuleger, E. (2022). Finding the Oldest Blockchain in the New York Times Classifieds. Retrieved from https://www.vice.com/en/article/j5nzx4/what-was-the-first-blockchain

Douceur, J.R. (2002). *The sybil attack*. Paper presented at the IPTPS.

Dwork, C. and Naor, M. (1992). *Pricing via processing or combatting junk meail*. Paper presented at the CRYPTO '92.

Else, H. (2022a). Papers co-authored by Nobel laureate raise concerns. *Nature* 611: 19–20.

Else, H. (2022b). What makes an undercover science sleuth tick? Fake-paper detective speaks out. *Nature* 608 (7923): 463–463.

Geim, A. (2010). Many pioneers in graphene discovery. *APS News* 19 (1): 4.

Hanson-Heine, M.W. and Ashmore, A.P. (2020). Computational chemistry experiments performed directly on a blockchain virtual computer. *Chemical Science* 11 (18): 4644–4647.

Hanson-Heine, M.W. and Ashmore, A.P. (2023). Blockchain technology in quantum chemistry: a tutorial review for running simulations on a blockchain. *International Journal of Quantum Chemistry* 123 (4): e27035.

Kaushik, R., Nawathean, P., Busza, A. et al. (2007). PER-TIM interactions with the photoreceptor cryptochrome mediate circadian temperature responses in drosophila. *PLoS Biology* 5 (6): e146. https://doi.org/10.1371/journal.pbio.0050146.

Kaushik, R., Nawathean, P., Busza, A. et al. (2016). Retraction: PER-TIM interactions with the photoreceptor cryptochrome mediate circadian temperature responses in drosophila. *PLoS Biology* 14 (2): e1002403. https://doi.org/10.1371/journal.pbio.1002403.

Kincaid, E. (2023). Nobel Prize winner Gregg Semenza tallies tenth retraction. Retrieved from https://retractionwatch.com/2023/10/02/nobel-prize-winner-gregg-semenza-tallies-tenth-retraction/

Lai, R., Pu, C., and Peng, X. (2018). On-surface reactions in the growth of high-quality CdSe nanocrystals in nonpolar solutions. *Journal of the American Chemical Society* 140 (29): 9174–9183.

Lamport, L., Shostak, R., and Pease, M. (2019). The Byzantine generals problem. In: *Concurrency: the works of leslie lamport*, pp. 203–226.

Lin, P.-Y., Chen, L.Y., Zhou, P. et al. (2023). Neurexin-2 restricts synapse numbers and restrains the presynaptic release probability by an alternative splicing-dependent mechanism. *PNAS* 120 (13): e2300363120.

Litovchick, A. and Szostak, J.W. (2008). Selection of cyclic peptide aptamers to HCV IRES RNA using mRNA display. *PNAS* 105 (40): 15293–15298. https://doi.org/10.1073/pnas.0805837105.

Litovchick, A. and Szostak, J.W. (2009). Retraction for Litovchick and Szostak. "Selection of cyclic peptide aptamers to HCV IRES RNA using mRNA display". *PNAS* 106 (17): 7263–7263. https://doi.org/10.1073/pnas.0903110106.

Network, C. (2018). The ASIC Resistance of Proof of Space. Retrieved from https://www.chia.net/2018/06/11/the-asic-resistance-of-proof-of-space/

Nakamoto, S. (2008). Bitcoin: a peer-to-peer electronic cash system. Satoshi Nakamoto.

Potti, A., Dressman, H.K., Bild, A. et al. (2006). Genomic signatures to guide the use of chemotherapeutics. *Nature Medicine* 12 (11): 1294–1300. https://doi.org/10.1038/nm1491.

Potti, A., Dressman, H.K., Bild, A. et al. (2011). Retraction note: genomic signatures to guide the use of chemotherapeutics. *Nature Medicine* 17 (1): 135–135. https://doi.org/10.1038/nm0111-135.

Van Noorden, R. (2023). More than 10,000 research papers were retracted in 2023—a new record. *Nature* 624 (7992): 479–481.

Wood, G. (2014). Ethereum: A secure decentralised generalised transaction ledger. *Ethereum Project Yellow Paper* 151 (2014): 1–32.

Xu, Y., Liu, R., Li, J. et al. (2020). The blockchain integrated automatic experiment platform (BiaeP). *Journal of Physical Chemistry Letters* 11 (23): 9995–10000.

Yamamizu, K., Iwasaki, M., Takakubo, H. et al. (2017). RETRACTED: in vitro modeling of blood-brain barrier with human iPSC-derived endothelial cells, pericytes, neurons, and astrocytes via notch signaling. *Stem Cell Reports* 8 (3): 634–647.

Yamamizu, K., Iwasaki, M., Takakubo, H. et al. (2018). Retraction notice to: in vitro modeling of blood-brain barrier with human iPSC-derived endothelial cells, pericytes, neurons, and astrocytes via notch signaling. *Stem Cell Reports* 10 (2): 674.

Zeng, M., Hu, Z., Shi, X. et al. (2014). MAVS, cGAS, and endogenous retroviruses in T-independent B cell responses. *Science* 346 (6216): 1486–1492.

7

The Future Integrated Computational and Experimental Research in Metaverse

7.1 Introduction of Metaverse

The metaverse concept, initially introduced by Neal Stephenson in "Snow Crash" (1992), has evolved into a complex digital universe that merges augmented reality (AR), virtual reality (VR), and the internet into a collective, immersive space for diverse activities like gaming, social interactions, education, and commerce. Central to its development are advancements in theory, hardware, and software, creating a user-inhabited universe that exceeds traditional digital experiences.

Crucially, blockchain technology underpins the metaverse, establishing a symbiotic relationship that enables decentralized governance, economy, and interaction within these virtual worlds. Blockchain's key features – decentralization, security, transparency, and interoperability – facilitate secure transactions, digital asset ownership, and user privacy. This infrastructure supports complex virtual economies, where digital currencies and non-fungible tokens (NFTs) allow users to transact, own unique digital items, and maintain secure digital identities. By leveraging blockchain, the metaverse promises a more integrated, authentic, and user-controlled digital experience, showcasing the technology's potential to reshape virtual spaces and economies profoundly.

7.1.1 Industry 5.0 Protocol

Industry 5.0 represents the next evolution in the industrial revolution, emphasizing the integration of advanced technologies with human ingenuity and creativity to achieve more sustainable, personalized, and efficient production processes. Unlike Industry 4.0, which focused on automation and data exchange in manufacturing technologies, Industry 5.0 places a strong emphasis on the collaboration between humans and machines, aiming to enhance human capabilities and ensure that technological progress benefits society as a whole. Metaverse technology plays a pivotal role in realizing the vision of Industry 5.0 by providing a framework for immersive, interactive, and interconnected virtual environments that can revolutionize various aspects of industry, including enhanced collaboration and remote work, training and skill development, digital twins and simulation, customized and consumer-centric product, and sustainable practices and circular economy.

AI and Robotic Technology in Materials and Chemistry Research, First Edition. Xi Zhu.
© 2025 WILEY-VCH GmbH. Published 2025 by WILEY-VCH GmbH.

Among these applications, the most promising aspect should be the integration of AI and robotics. The metaverse serves as a platform for integrating AI and robotics into industrial processes in a way that complements human workers rather than replacing them. This symbiotic relationship maximizes efficiency and innovation while ensuring that technology serves to augment human capabilities. By simulating and optimizing production processes in the metaverse, companies can identify ways to reduce waste, improve energy efficiency, and adopt more sustainable practices. Meanwhile, industries can leverage digital twins (i.e. virtual replicas of physical systems or processes) to simulate and analyze performance in real time. This aids in optimizing operations, predicting maintenance needs, and reducing downtime, leading to more efficient and sustainable practices. We will discuss this part in detail in the context of physics and chemistry research in the following sections.

7.1.2 Current Development of Metaverse

The recent decades have witnessed the rapid development of metaverse in both hardware and software. The hardware necessary for the metaverse encompasses a wide array of devices designed to facilitate immersive experiences. This includes VR & AR headsets, wearable technology and high-performance computing devices. Devices such as the Oculus Rift, HTC Vive, PlayStation VR, and Microsoft HoloLens that offer immersive visual and auditory experiences, placing users inside the virtual world. Besides, haptic feedback suits, gloves, and even treadmills that enable users to experience tactile sensations and move naturally in the virtual space, enhancing the sense of immersion. The Apple Vision Pro, launched on February 2, 2024, has generated a mix of excitement and scrutiny among users and reviewers alike. Priced starting at $3,499 in the United States, it represents Apple's ambitious foray into the mixed reality space, heralded as a significant leap toward spatial computing. Interaction with the device is predominantly through eyes, hands, and voice controls, offering a new layer of organic interaction within the digital environment. Reviewers have praised its hardware and immersive experience, particularly high-lighting the quality of its displays and the spatial audio features. The establishment of metaverse is based on the solid support of high-performance computing devices. Powerful PCs, gaming consoles, and dedicated servers are crucial for rendering high-fidelity graphics, processing complex interactions in real time, and supporting the vast, persistent worlds of the metaverse.

The software driving the metaverse includes a broad ecosystem of platforms, development tools, and protocols that enable the creation, maintenance, and exploration of vast virtual spaces. Key components include platforms and engines, networking and infrastructure, and content creation tools. Software like Unity and Unreal Engine allows developers to create complex, interactive 3D environments. Platforms such as Roblox, Fortnite, and Decentraland offer ready-made spaces for users to explore, socialize, and create. Crucial software infrastructure ensures that these virtual worlds are accessible, scalable, and secure. This includes cloud computing services, blockchain technologies for digital ownership and transactions, and protocols for interoperability between different virtual spaces. Software tools enable users and developers to create digital assets, design interactive experiences, and script complex behaviors within the metaverse. This also includes AI-driven tools that can generate dynamic content or assist in designing virtual environments.

7.1.3 Human-in-Loop Paradigm

Human-in-loop (HIL) optimization integrates human expertise into automated systems, enhancing their performance, accuracy, and adaptability. This approach is crucial in Industry 5.0 and the metaverse, where human judgment refines content moderation, improves virtual design, and trains AI models by correcting and refining their outputs. Foldit and AlphaFold exemplify HIL's impact in scientific research. Foldit, an online game launched in 2008, demonstrates the power of human intuition in protein folding, contributing to significant scientific discoveries, such as deciphering an AIDS-related enzyme structure. AlphaFold, developed by DeepMind, leverages deep learning to predict protein structures with unprecedented accuracy, benefiting from human expertise in its algorithm design. These cases illustrate the complementary strengths of human intelligence and AI in solving complex problems and advancing scientific knowledge.

The collaboration between humans and machines accelerates scientific discovery and makes participation in research more accessible, highlighting the potential for more inclusive and innovative scientific processes. However, these advancements underscore the reliance on computational data, suggesting a role for AI-embedded metaverse platforms in bridging the gap between virtual models and experimental validation, potentially expanding the scope and application of scientific research in virtual environments.

7.2 Research Paradigm in Metaverse

Paradigm shift originates from the development of production tools. The traditional research diagram emphasizes theory over data, where theory is established based on the analysis of experiment data through human intelligence (Hoffmann and Malrieu 2020). Scientists used computer simulation to construct theoretical calculation methods. In traditional research paradigm, normal experiments are analyzed with theory and simulations, then researchers aim to transfer the simulation result to new experiments, which generally fails. This is illustrated in Figure 7.1a. With the evolution of computational power and technology, the generation of new data and the transferability of models, driven by AI, are increasingly valued over mere accuracy (Back et al. 2024). Researchers are now ambitiously aiming to scan the entire material space (Merchant et al. 2023; Park et al. 2024), or develop comprehensive algorithms capable of computing entire chemical systems (Batatia et al. 2023; Firaha et al. 2023). To effectively interact with this burgeoning data, a sophisticated interface is imperative. The metaverse offers an ideal platform for this digitalization process, seamlessly integrating data, theory, and human intelligence. Within this framework, experimental data is harmonized with theoretical models

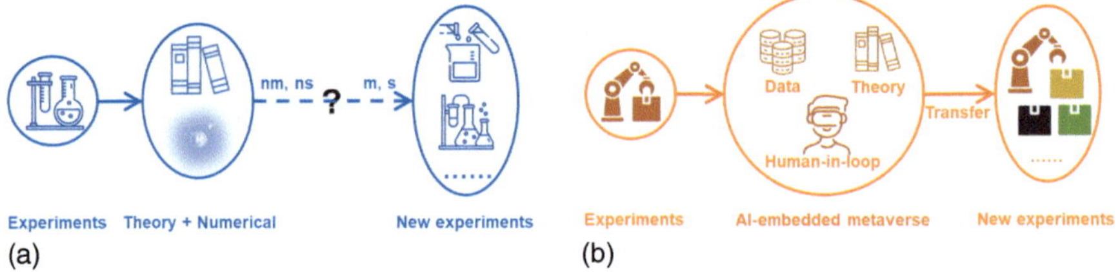

Figure 7.1 Comparison of the research paradigm between the conventional paradigm (a) and its digitalized counterparts (b).

and AI algorithms. Through human-in-the-loop optimization processes(Wu et al. 2022), both theoretical and AI models are refined efficiently within the metaverse, enabling humans to circumvent the pitfalls of local minima or overfitting. For instance, while traditional methods require awaiting the completion of a chemical reaction for yield calculation and backpropagation, the metaverse allows for the direct optimization of reaction parameters, leveraging simulated data on the process. Unlike conventional simulations, outcomes calculated within the metaverse encompass comprehensive time and space scales, allowing for the direct prediction of unknown experimental states. As optimization processes utilize increasing volumes of data, both the accuracy and transferability of the models are enhanced. This novel research paradigm is depicted in Figure 7.1b.

In the AI-embedded metaverse research paradigm mentioned in Figure 7.1, there are several key components, which will be detailly illustrated part by part: autonomous high-throughput experiments, theory driven by AI, HIL implementation and AI predictions.

7.3 Autonomous High-Throughput Experiments

In recent decades, a number of robotics systems have been implemented to do physics and chemistry experiments, ranging from universities and industries. Robotics systems can execute tasks with high precision and consistency, reducing human error in experiments. This is crucial in physics and chemistry, where exact measurements and consistent conditions are vital for valid results. Compared to conventional localized experiments, robotics system features high throughput, remote operations, and automation, facilitating in situ data collection and analysis. The data is delicately integrated and input to metaverse. The users can monitor numerous hardware devices instead of one. To improve the compatibility the metaverse to connect the hardware as many as possible, the Materials Acceleration Operation System (MAOS) is proposed (Li et al. 2020). The synthesis methods, as well as experiment operation, amounts of chemical species are critically parametrized into geometry symbols and mathematical notations.

In the current era, the advent of digitalization and its transformative impact on the research paradigm, especially in chemical experiments, is increasingly evident. Traditionally, conducting numerous experiments was necessary to obtain results, which were then individually analyzed to hypothesize about the optimal reaction conditions for superior product yields. This method, while useful, often resulted in approximations that lacked precision. With the introduction of digitalization, a shift toward a data-driven methodology allows for the setting of specific threshold values for desired properties of reaction products. The premise is that products meeting these threshold criteria are considered optimal. Should these criteria not be met, an AI-driven "black box" undertakes the task of continuous optimization toward these predefined standards. The question arises: is there an inherent logic to this process? While the necessity for deep chemical reasoning may seem diminished, the integration of AI and digital tools streamlines the experimental and operational processes into a unified digital format. This evolution prompts a reevaluation of our approach to enhancing experimental data through the digitization of laboratory environments. When an AI-equipped robotic system operates within a standardized experimental protocol for over 10,000 hours, gathering data and refining models, it earns a level of reliability comparable to the industrial automation seen in enterprises like SIEMENS. This paradigm illustrates a future where AI and robotics systems, trusted through extensive training and data accumulation, revolutionize chemical experimentation.

To optimize the human–AI interface in the metaverse, our framework ensures realistic user experiences by aligning digital observation and temporal scales with real-world counterparts. It comprises a digital interface and a backend tensor computation module. This module computes system states at the atomic level over extensive time scales, while the interface visualizes these results on macroscopic and microscopic scales. Users experience a virtual chemistry lab, utilizing high-throughput data to refine backend computations for tasks like crystallography and drug design. Our proposed framework, illustrated in Figure 7.2, includes user terminals, cloud servers, and remote laboratories. Time-sensitive experimental

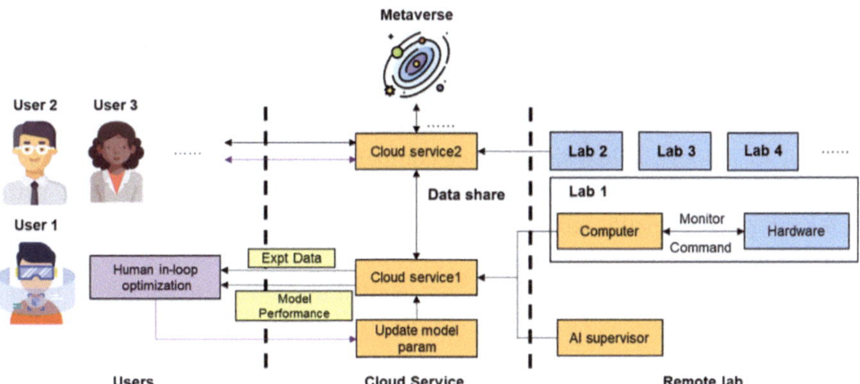

Figure 7.2 Detail framework unifying human-in-loop optimization and experiment data. There are three regions representing user terminals (purple boxes), cloud servers (orange boxes), and remote laboratories (blue boxes). The time-dependent experiment data will be generated and recorded in the remote lab and uploaded to cloud services. The cloud service is the place for high-performance calculation, data optimization, and distribution centers. The connections between different services are designed to guarantee data share.

data, generated in remote labs, is uploaded to cloud services for high-performance calculations and data optimization. This setup ensures efficient data sharing. User 1 receives experimental and model performance data, aiding in model optimization. Modified parameters are updated in the cloud, facilitating human-in-loop optimization in the metaverse. Compared to conventional algorithms, humans have a deeper understanding of the correlations in the atomic system; thus, they can interfere with the computer to escape the local minimum. Users access data from journals, databases, and real-time experiments. Recent advancements introduce an AI supervisor for literature and data analysis.

7.3.1 Theory Driven by AI

Current ab initio methods such as density functional theory (DFT) (Makkar and Ghosh 2021) and molecular dynamics (MD) (Hollingsworth and Dror 2018) have obvious deficiencies. DFT can only calculate the system properties at 0K. For classical MD, the system is adiabatic, and the time step is far from bulk. It takes about a day for the latest supercomputer to calculate 100 ms simulation in the field of biology (Shaw et al. 2021). It seems meaningless for users to inspect an intermediate state of femtosecond or nanosecond in an organic reaction taking several hours. Even so, there are numerous phenomena that humans still cannot explain with the current force field model, especially in life science (Karplus and McCammon 2002). Therefore, the simulation result cannot resemble the real experiment counterparts in both space and time scales, which is labeled as the lower dashed line. The conventional research diagram cannot form a closed loop.

Therefore, in the metaverse, the theory should be driven by AI instead of chemical calculations. This can be illustrated from two perspectives. Firstly, the knowledge can be delivered from large language models AI such as AI-supervisor and Spider Matrix. Through conversation with AI, users can gain knowledges from journal article analysis. Secondly, we believe that the researchers can bypass the specific physics and chemistry knowledge, directly receiving information from in situ experiment data with AI analysis. How do we guarantee the accuracy of AI? Concerns may rise that there is lack of explainability of AI in real physics world. In fact, the explainability of AI model has less significance compared to the transferability of model. Users do not care if they can understand the model, whereas they do care if the model can predict results accurately and if the model can be advocated to a wide range of industries.

Then, we will illustrate a possible AI-embedded framework to substitute for the traditional chemistry calculations. The *in situ* experiment data may consist of temperature, pressure, Raman spectroscopy, and high-performance liquid chromatography (HPLC). From initial experiments recipe, users are able to know the amount of initial chemical species and the volume of reactor. Here the first AI model will be established to analysis the spectrum data to obtain the current system configuration. In the Raman spectrum, the position of peak determines the specific functional group, while the full width at half maximum (FWHM) of the peak reveals the abundance of the species. Similarly, the HPLC reveals system configuration information.

By analyzing this information, AI can calculate several possible system configurations based on the initial recipe. From this system configuration, an atomic trajectory can be generated. From the system configuration as well as the atomic trajectory, the users will judge whether the reaction proceeds normally, or the experiments parameters are appropriate. This will be illustrated in the HIL Implementation part in the following section.

7.3.2 HIL Implementation

In the initial stages of AI application development, human instructions are converted into codes or mathematical inputs through a process known as coding, which computers can then compile and execute. This process represents an "Indirect interaction" between human intelligence and research problems, as depicted in Figure 7.3a. To align with Industry 4.0's emphasis on digitalizing human intelligence, entities such as Neuralink have explored the establishment of brain–machine interfaces to address this challenge (Abiri et al. 2019; Musk 2019; Rezeika et al. 2018), though this area is fraught with ongoing medical and ethical debates. Conversely, the emergence of metaverse technology offers a novel approach. Rather than serving merely as a "coding machine," computers could function as intermediaries between human intelligence and data, facilitating a "direct interaction." This model allows human insights, including knowledge of data and its correlations, to significantly influence model optimization processes, as illustrated in the lower flow of Figure 7.3a.

The prevailing approach in MD simulations adheres to the indirect interaction model, as shown in Figure 7.3b. Various computational frameworks, such as ab initio MD (Car and Parrinello 1985) and neural-network-based methods (Zhang et al. 2018), compute system properties in a digital format, relying on digitized physical chemistry information. However, accurately digitalizing the intricate interactions due to complex correlations remains a challenge. For both ab initio MD and deep-learning-based MD approaches (Yann et al. 2015, 1988), the emphasis tends to be on data rather than on data correlation, leading to a decoupling of feature properties. This results in a feature space represented as a one-dimensional vector, lacking in detailed correlations, as illustrated in Figure 7.3b.

Metaverse technology introduces a paradigm shift in the computation of feature correlations, especially once a human researcher engages with the system. Through the metaverse, researchers can observe and analyze trajectories, allowing them to assess the accuracy of the force field based on their expertise. This process enables the translation of property correlations into a two-dimensional force field matrix. This matrix represents a significant advancement over the traditional one-dimensional force field parameters typically used in conventional MD simulations. The optimization process facilitated by the metaverse is illustrated in Figure 7.3c, where the properties of original materials are depicted as boxes of various colors. Researchers have the capability to fine-tune the force field matrix correlations by examining the resultant properties, conducting this qualitative analysis mentally. To validate the efficacy of this correlation mapping, they can

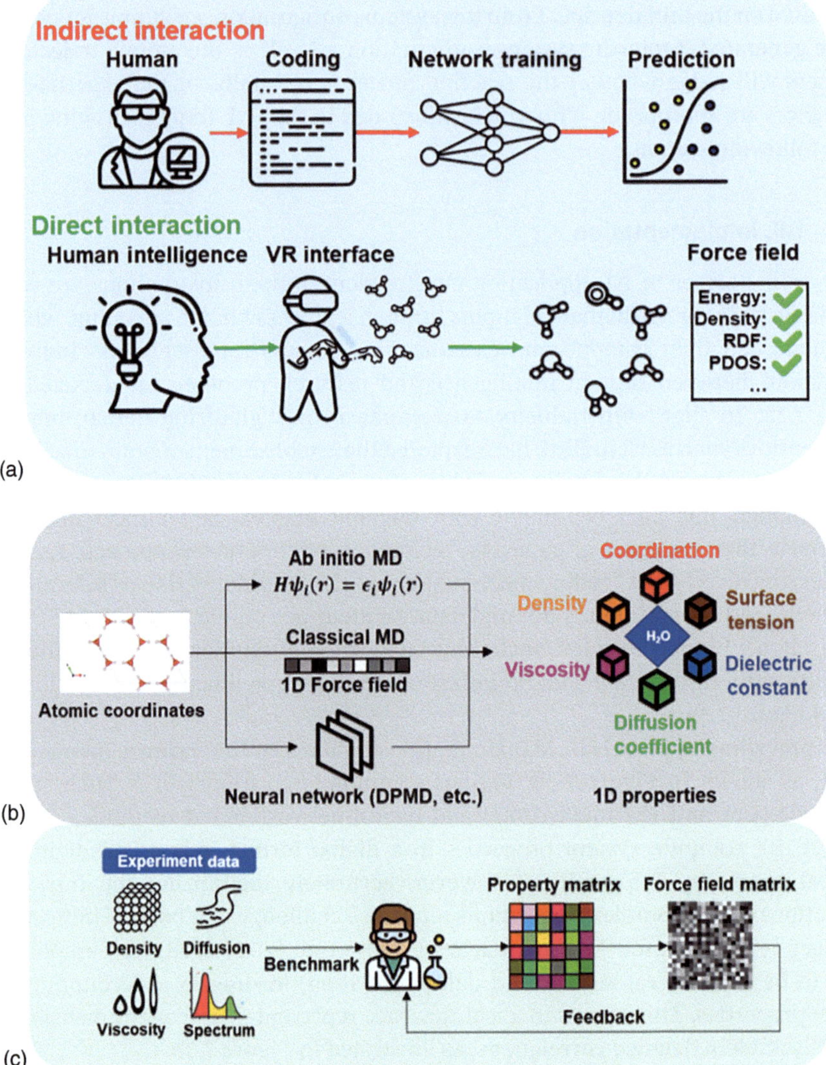

Figure 7.3 (a) Comparisons of current calculation paradigm and the VR-based direct interaction counterpart. (b) Current calculation scheme of MD simulations, where properties are output as 1D vector from the 1D force field. (c) Dimension increases the force field by endorsing the correlation force field matrix and correlation storage of the human brain.

initiate an MD simulation using the refined force field vector. This simulation generates a new set of correlated properties, thereby creating a comprehensive closed-loop optimization process. The metaverse environment supports the execution of multiple MD simulations, incorporating all relevant parameters (such as force field parameters and neural network parameters) into an extensive correlation matrix. This matrix not only enhances the depth of analysis but also facilitates a more nuanced understanding of the material properties and their interactions.

Through this advanced approach, the metaverse technology empowers researchers to conduct more accurate and insightful simulations, pushing the boundaries of material science research.

7.3.3 AI Prediction

After manipulating the correlation matrix/tensor, we want to optimize an AI model to achieve two goals. The first goal is to predict the results of the training experiment ahead of time. In this way, the users can avoid waiting for a bad experiment result and adjust experiment parameters accordingly. The second goal is to transfer the AI model to various experiments.

In our new work to achieve the first goal, our project is focused on simulating physical chemistry processes, utilizing a U-net architecture (Ronneberger et al. 2015) and the variational autoencoder (VAE) (Kingma and Welling 2013), to achieve a comprehensive digital representation of chemical phenomena. This approach is designed to augment, rather than replace, traditional MD simulations by incorporating both microscopic and macroscopic perspectives into the computational analysis. Unlike standard diffusion models that primarily encode hyperparameters, our model is specifically tailored to encode actual physical properties, notably the force field parameters, directly into the network's computations. This inclusion of force field parameters significantly enhances the model's applicability and transferability across various chemical systems. The process begins with inputs such as chemical species, quantities, temperature, and pressure, as elaborated in Figure 7.4. The model employs a VAE to integrate these inputs and reconstruct the expected system state, complemented by a diffusion-like mechanism facilitated by a U-net structure. The inputs comprise the initial atomic trajectory and an overall view of the system's bulk appearance. Atomic trajectories are transformed into images, with their resolution determined by the molecule count, while the bulk system appearance is encoded into a latent variable, Z, through the VAE. This encoding effectively captures the complete state of the system. Subsequently, a diffusion-like

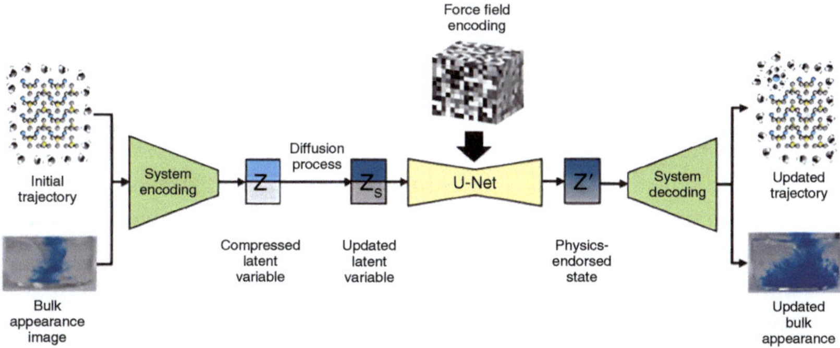

Figure 7.4 Network architecture of the physics-endorsed diffusion-like model.

process extrapolates the future state of Z, utilizing force field parameters to guide the projection.

This methodology underscores the dynamic relationship between the detailed micro-level interactions and the overarching macroscopic properties of the material, illustrating how atomic and molecular behaviors directly influence the observable characteristics of materials. As a result, the model generates a sequence of updated images reflecting changes at both the atomic/molecular and bulk material scales. These images are then iterated upon, creating a time series that depicts the chemical process evolution over time. The intervals between these updates can be adjusted to align with human perception capabilities, thereby enabling the estimation of the duration of physical chemistry processes under specified hardware conditions.

7.4 H_2O Phase Research in Metaverse

All the above sections describe the how experiments and AI are integrated in the metaverse. We are going to provide some examples in inorganic chemistry to show the advantages of metaverse over traditional research paradigm.

The study of ice crystal polymorphs, numbering nineteen under specific conditions (Del Rosso et al. 2020; Gasser et al. 2021; Millot et al. 2019; Salzmann et al. 2021), offers critical insights into the hydrogen bond force field (Ignatov and Mosin 2014). Research on various force field models aims to capture water and ice's essential properties, such as mass density and dielectric constant (Kadaoluwa Pathirannahalage et al. 2021). These models are evaluated for accuracy against experimental data, with findings represented in a visual format indicating both accuracy and deviation (Kadaoluwa Pathirannahalage et al. 2021). Further examination reveals that while TIP3P force fields are widely used in biology, they show lower accuracy for H_2O properties compared to TIP4P force fields, which perform better for ice due to a focus on electron charges (Abascal et al. 2005). TIP5P models, considering lone pair electrons, excel in calculating dielectric properties (Khalak et al. 2018). Despite these advances, achieving broad accuracy across all water properties remains elusive due to the inherent trade-offs between long-range and short-range interactions, a dilemma compounded by the principles of quantum mechanics and the Heisenberg uncertainty principle. ·

Recent progress in water force field development faces challenges with transferability and the constraints of a low-dimensional computational approach. The evolution from the ab initio method to one-dimensional vector representations in MD simulations (Jorgensen et al. 1983) to more sophisticated matrix formulations signifies a move toward capturing correlations between data points. Anticipation grows for tensor computation to advance this field further by leveraging complex correlation information, marking a potential paradigm shift in computational chemistry.

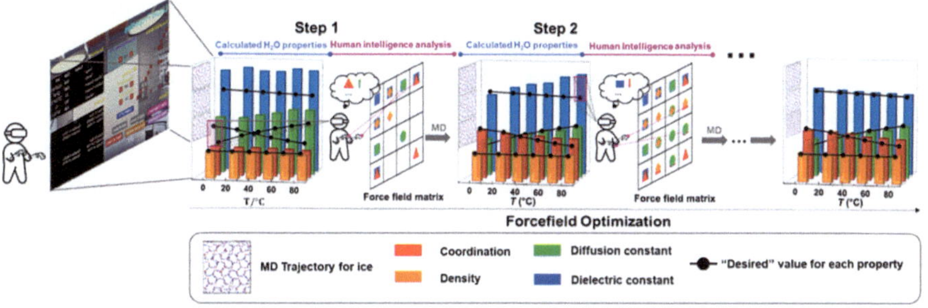

Figure 7.5 General diagram of HIL optimization in water force field. An individual optimizes the force field matrix according to the following criteria: fluid properties such as density, coordination number, diffusion and dielectric properties, and ice geometry structure through MD trajectory. After MD calculation, the player observes all the calculated properties through the current force field; then, he adjusts the corresponding diagonal and off-diagonal terms using his knowledge and understanding of the water system.

The metaverse operates on a principle that leverages human intelligence, particularly drawing from prior knowledge and the ability to store and utilize correlations. This approach mirrors the human problem-solving process, which involves collecting multi-dimensional information as inputs, applying logical thinking and knowledge to analyze these inputs, and producing outputs such as judgments, predictions, or actions. Human cognitive processes, akin to the operations of a real neural network whose structure remains largely undefined, serve as an analog for understanding how metaverse engages players in its interactive environment.

In the metaverse context, particularly for tasks requiring an understanding of physical chemistry, players are expected to have foundational knowledge that aids in identifying and leveraging correlations within a water system. This setup is depicted in Figure 7.5, illustrating the general workflow within the metaverse. Players are given the capability to adjust their force field matrix, which is represented in the user interface as a lower triangular matrix. This adjustment is part of the simulation setup, alongside setting conditions such as temperature and pressure, before proceeding to the computational phase. The metaverse is designed to support various computational backends, with LAMMPS currently being the computational software of choice.

Players engage with the simulation by observing properties and trajectories within the metaverse, guided by externally provided experimental values – the "desired" values they aim to achieve, highlighted by a yellow bar in each 3D plot. This observation phase prompts players to analyze why the current force field exhibits low accuracy, directing attention to properties that significantly deviate from desired outcomes. The interface aids in identifying correlations between different properties, emphasizing the importance of off-diagonal terms in the force field matrix over the conventional diagonal terms.

As players iterate through the optimization process, their understanding of the underlying correlations deepens, leading to a progressively filled correlation matrix and improved accuracy of the force field. This interactive and iterative process not only enhances the players' grasp of physical chemistry concepts but also illustrates the potential of human-augmented computational models in solving complex scientific problems.

We aim to optimize a water force field TIP4P-Meta by considering the correlation between the hydrogen bond interactions and non-rigid bonding interactions. The total energy can be expressed in the following basis:

$$E_{tol}^{TIP4P\text{-}Meta}(r, \theta, \theta_{hb}) = E_{coul}(r; q_H) + E_{LJ}(r; \varepsilon_{LJ}, \sigma_{LJ}) + E_{hb}(r, \theta_{hb}; \varepsilon_{hb}, \sigma_{hb}, n)$$
$$+ E_{bond}(r; k_r) + E_{angle}(\theta; k_\theta) \tag{7.1}$$

To better describe the hydrogen bond interaction, which is not explicitly contained in the original TIP force field model, we adopt E_{hb} from the DREIDING (Mayo et al. 1990) potential. If the force field are defined in conventional scalar form, the force field parameters are expressed as a one-dimensional column vector, such as f^{TIP4P} in Eq. (7.2). Considering the correlations between the force field parameters, TIP4P-Meta will be expressed as a matrix F, where the off-diagonal terms $F_{ij, i \neq j}$ represent the correlation between parameters:

$$f^{TIP4P} = \begin{pmatrix} q_H \\ q_O \\ \varepsilon_{LJ}^{OO} \\ \sigma_{LJ}^{OO} \end{pmatrix}, \quad F^{TIP4P-Meta} = \begin{pmatrix} q_H & F_{1,2} & F_{1,3} & \cdots & F_{1,12} \\ F_{2,1} & \varepsilon_{LJ}^{OO} & F_{2,3} & \cdots & F_{2,12} \\ F_{3,1} & F_{3,2} & \sigma_{LJ}^{OO} & \cdots & F_{3,12} \\ \vdots & \vdots & \vdots & \ddots & \vdots \\ F_{12,1} & F_{12,2} & F_{12,3} & \cdots & n \end{pmatrix} \tag{7.2}$$

Since the correlation between force field parameters is implicit, it is appropriate for the human player to participate in the optimization process. There are 12 tunable parameters: $k_r, k_\theta, \sigma_{LJ}^{OO,OH,HH}, \varepsilon_{LJ}^{OO,OH,HH}, q_H, \sigma_{hb}, \varepsilon_{hb}, n$, indicating F a 12×12 matrix, which has 66 pairs of correlation. The optimization of force field matrix F is the key to the global minimum of total energy since it balances all the local minimum for individual energy basis. The optimization procedure is described in Figure 7.5. The optimization process in the metaverse starts with standardizing original TIP4P force field parameters into a diagonal matrix. Diagonal terms are initialized at 1, representing the force field parameters, while off-diagonal terms, left blank initially, signify correlations yet to be identified. Players adjust these off-diagonal terms, engaging in a process that mirrors deep-learning mechanisms, but with a unique emphasis on human intelligence to overcome local minima and enhance the model with complex, high-dimensional data about the atomic system. As shown in Figure 7.5, adjustments in the force field matrix F are projected onto a radial information tensor R, which then informs the backend computation and simulation processes, determining the properties of the water system.

To streamline the optimization, participants are advised to calibrate the force field closer to existing models like TIP4P, using experimental values across different temperatures as benchmarks (Harris and Woolf 1980; Jorgensen and Madura 1985; Malmberg and Maryott 1956). This iterative refinement, requiring approximately 50 attempts, leads to significantly enhanced accuracy. The resultant force field TIP4P-Meta exhibits improved predictions of liquid properties across temperatures.

This includes calculations of coordination numbers, density, dielectric, and diffusion properties at various temperatures, where TIP4P-Meta shows commendable accuracy. Furthermore, the model excels in predicting melting and dielectric properties of water, with the melting point of Ice Ih and the dielectric constant closely aligning with experimental values, demonstrating the efficacy of leveraging human insight in computational model optimization.

We also used TIP4P-Meta to calculate the density for different ice polymorphs and the phase diagram benchmarked with the TIP3P, TIP4P, and TIP5P models. Since MD in ultra-high pressure are unstable, ice X (Polian and Grimsditch 1984) and XVIII (Millot et al. 2019) are excluded. The result shows that TIP4P-Meta remains the accuracy of TIP4P/2005 in ice polymorph density calculation. Notably, for ice XVI, an ice structure with the lowest density of about 0.81 at negative pressure conditions, previous force fields all yield large deviation, while an approximate density of 0.79 is obtained through TIP4P-Meta. The ice phase diagram is computed using Gibbs–Duhem integration methods(Kofke 1993). The solved phase diagram is shown in Figure 7.6, compared with other TIP force fields. It is transparent to see that TIP4P-Meta further meets the experimental data, which also agrees with the accurate ice Ih melting point prediction.

To evaluate the transferability of our force field model, we conducted two demonstration applications: exploring new ice polymorphs and calculating the solvation energy of organic molecules. These applications are pivotal in crystal

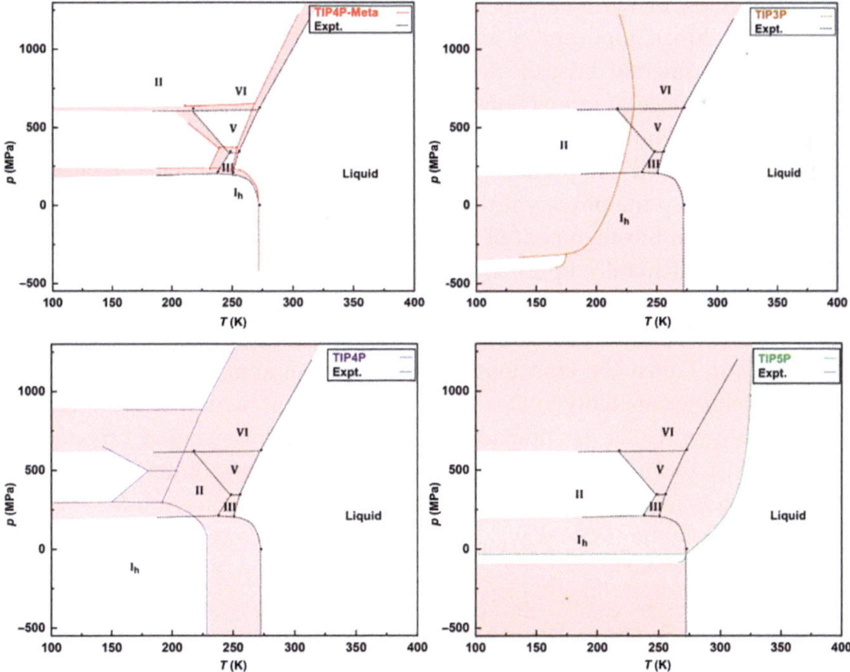

Figure 7.6 Ice phase diagram for TIP4P-Meta compared with original TIP3P, TIP4P, and TIP5P models.

polymorph design and the pharmaceutical industry, as both crystal formation and solubility are crucially dependent on molecular force fields. Inadequate force field models can lead to the development of unwanted crystal forms with incorrect structures or properties, potentially causing commercial setbacks and financial losses. A notable instance involved Abbott Laboratories having to reformulate the anti-HIV drug Ritonavir due to the emergence of a more stable polymorph during production, which had a different molecular conformation (Bauer et al. 2001). Similarly, a new polymorph of aspirin was discovered, highlighting the critical role of hydrogen bond geometry and intramolecular forces (Li 2007; Shtukenberg et al. 2017).

In our first task, we applied TIP4P-Meta to discover new forms of ice polymorphism, adjusting water molecule counts from 12 to 128 to generate a wide range of candidates with USPEX (Glass et al. 2006; Lyakhov et al. 2013). After selecting the most stable structures, we further refined them in the metaverse, allowing for an immersive inspection and optimization process. Stability assessments, including phonon density of states (PDOS) analyses, confirmed the meta-stability of these structures compared to ice Ih under negative pressure.

For the second task, we focused on hydrogen bonding in organic solutions at ambient conditions, employing TIP4P-Meta for water and CHARMM parameters for organic molecules, as generated by SwissParam Server(Zoete et al. 2011). We assessed solvation-free energy to evaluate solute–solvent interactions, underlining the water force field's significance in solvation processes. We tested a range of organic molecules, including ethane, benzene, urea, and aspirin, to demonstrate our force field's broad applicability and transferability. The computed results align closely with experimental data, showcasing the optimized force field's effectiveness across different molecular interactions.

The distinction of human intelligence compared to AI or ML models lies significantly in the nuanced comprehension of nature. The capacity of deep-learning models to fully grasp the physics at the electron scale is a subject of debate, partly due to the potential misalignment of test data (Gerasimov et al. 2022; Kirkpatrick et al. 2021, 2022). Generally, optimization strategies involve a linear amalgamation of all conceivable mathematical bases. The linear coefficient serves dual purposes: firstly, to activate or deactivate a basis via assigning a value of zero or non-zero, and secondly, to adjust the contribution to the overall sum. Often, optimization algorithms lack the capability to toggle the basis on or off, which can lead to computational failures. However, humans, overseeing the optimization process, can manually adjust the basis as needed. This synergy between human insight and ML models tends to yield superior outcomes than relying solely on ML models, independent of computational power.

7.5 Aqueous System Research in Metaverse

In comparison to pure water, the presence of ions introduces distinct properties to ionic solutions due to the intricate correlations between ions and water molecules.

Traditional force fields, such as Madrid-2019 (Zeron et al. 2019), AZ-FF (Broo and Nilsson Lill 2016), and DREIDING (Mayo et al. 1990), predominantly model the aqueous system in its steady state, assuming chemical bonds to be static. This approach is inadequate for capturing dynamic processes like hydrolysis, or the states of boiling and freezing. Reactive force fields might offer a solution, yet a notable gap exists in the correlation between force field parameters, impacting their transferability – a challenge we highlighted in our prior research (Gao et al. 2022). To address this, we have aggregated almost all reactive force fields (Aryanpour et al. 2010; Fedkin et al. 2019; Monti et al. 2013; Rahaman et al. 2010) pertinent to aqueous systems and proposed expanding the conventional vector-like force field parameters into a comprehensive force field matrix. This matrix, which does not necessarily adhere to symmetry, aims to encapsulate a broader spectrum of interactions, thereby enhancing the descriptive power and transferability of the force field for modeling dynamic aqueous system:

$$
F_{ij} = \begin{bmatrix} F_{\mathbf{bond}} & \sigma_{12} & \sigma_{13} & \sigma_{14} \\ \sigma_{21} & F_{\mathbf{vdWaals}} & \sigma_{23} & \sigma_{24} \\ \sigma_{31} & \sigma_{32} & F_{\mathbf{Coulomb}} & \sigma_{34} \\ \sigma_{41} & \sigma_{42} & \sigma_{43} & F_{\mathbf{hb}} \end{bmatrix}
\tag{7.3}
$$

where $F_{\mathbf{bond}}$, $F_{\mathbf{vdWaals}}$, $F_{\mathbf{Coulomb}}$, $F_{\mathbf{hb}}$ denotes the block force field matrix of chemical bond, van der Waals interaction, Coulomb interaction, and hydrogen bond. The enhancement of reactive force field parameters involves strategic placement in a matrix format, where diagonal terms are reserved for original parameters and off-diagonal terms represent their interrelations. The block matrix σ_{ij} on off-diagonal positions signifies correlations between distinct parameter blocks, facilitating a more interconnected and "correlated" force field parameterization, subsequently diagonalized from F_{ij}. This complex force field matrix is derived from a three-dimensional radial information tensor, itself expanded polynomially from empirical data fitting of molecular force field matrices against stable atomic trajectories.

Figure 7.7 illustrates the iterative optimization of the force field tensor within a user interface that displays atomic trajectories and current properties for evaluation against experimental benchmarks. Users can adjust hyper-parameters in real time

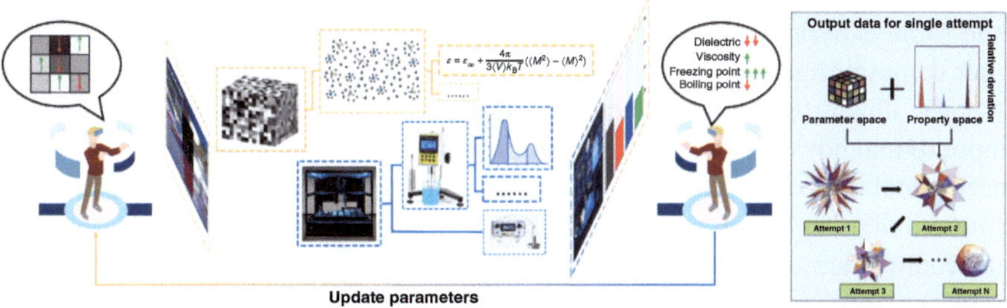

Figure 7.7 Implementation of physics-endorsed model and human-in-loop force field tensor optimization scheme. Left figure: Human operation in single optimization iteration. Right figure: The optimization procedure as the function of human attempts.

to streamline the optimization process. The iteration outputs, including parameter adjustments and property accuracy, are categorized with deviations labeled as "spikes" and parameters as the "core," facilitating targeted refinements through human-in-loop optimization. This process effectively reduces initial deviations, as demonstrated in the optimization and simulation of copper sulfate pentahydrate solvation, supported by video datasets of the solvation process under varied shaking speeds for comprehensive training.

Using a real-time physics-endorsed model, our experiments with NaCl and $CuSO_4 \cdot 5H_2O$ were conducted with Fine-Fanta's specialized hardware. These bulk appearance changes were recorded by a camera on the hardware. The hardware also autonomously collected data on properties like density and pH for metaverse analysis at regular intervals.

Figure 7.8a illustrates the timeline of the $CuSO_4$ solvation process, comparing real-world events to predictions made by a metaverse-based model. Initially, the model's forecasts, depicted by a dashed line, trail behind the actual solvation events, symbolized by a solid line. However, as the model adjusts, it starts to anticipate future developments, eventually predicting events ahead of the real-time process, as highlighted by the blue-shaded area. The dissolution of $10\,g\ CuSO_4 \cdot 5H_2O$ in 100 ml deionized water, stirred at 300 r/min, is observed to take approximately 15 minutes. For model training within the metaverse lab, data from the first four minutes of this experiment, including specific hardware settings, are utilized. Subsequently, the model demonstrates the capability to simulate the entire dissolution process in mere seconds, accurately forecasting intermediate and final states. To assess the model's robustness, we explored various solvation conditions, altering temperature ($25–50\,°C$) and stirring power ($100–500\,r/min$), confirming the model's reliability across these varied settings.

Figure 7.8b explores the impact of computational power on accelerating physical chemistry processes, analyzing both traditional experiments and their digital twins within the metaverse across three dimensions: time consumption, computational power utilization, and data generation volume. Traditional laboratory tasks, such as adding solvents or shaking mixtures, exhibit moderate time consumption, minimal data generation, and no reliance on computational resources. Conversely, metaverse-based optimization processes demand high computational power and generate substantial data volumes. When comparing two metaverse simulations of the same experiment, differing in computational power usage, it's shown that higher computational power can significantly reduce the time required to predict outcomes, illustrating an "acceleration" effect. For instance, a model with greater computational power might predict a 15-minute reaction based on just 2 minutes of experimental data. This efficiency is poised to increase with advancements in computing paradigms, like quantum computing, suggesting that experiments taking hours could be simulated in the metaverse in mere seconds, thereby revolutionizing how scientific investigations are conducted.

After successfully simulating the dissolution processes and calculating the physicochemical properties of $CuSO_4$ and NaCl solutions across all temperatures and concentrations, generating 800 data entries (metadata), we sought to extend our

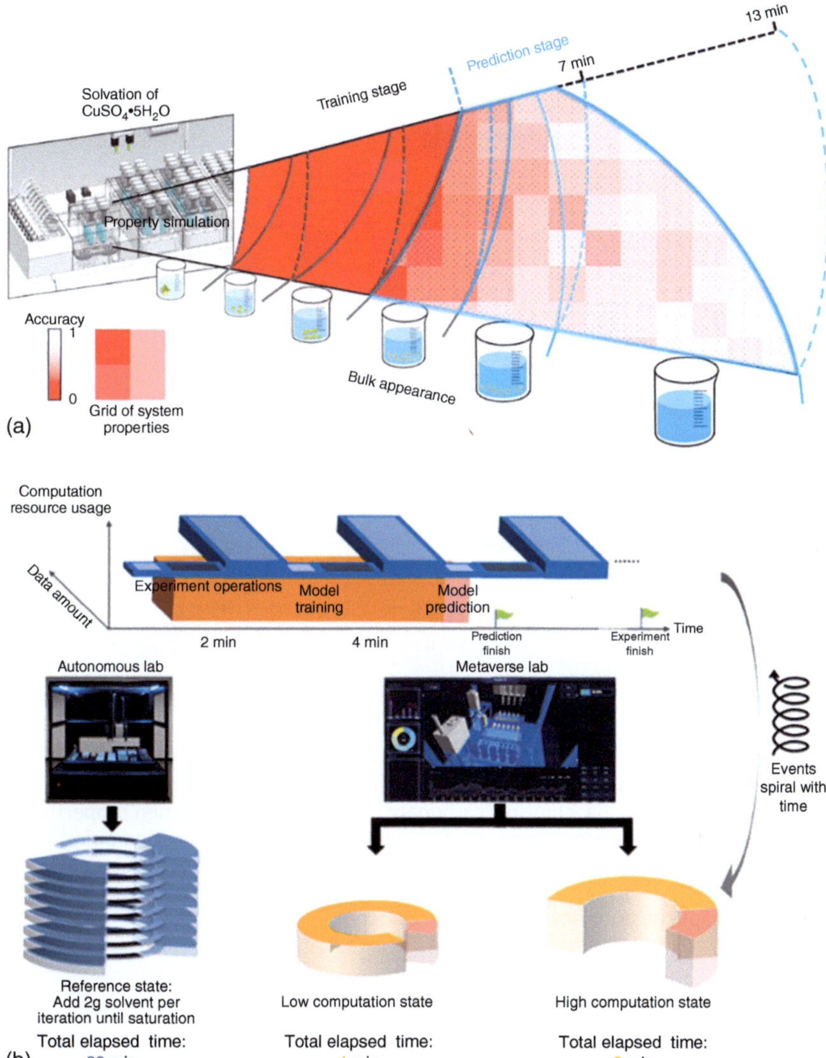

Figure 7.8 Solvation prediction of $CuSO_4$ in the metaverse lab. (a) The diagram showcases the $CuSO_4$ solvation process prediction. At the bottom, the figure captures the bulk appearance observed during the experiment. Central to the figure is a grid that assesses the model's predictive accuracy regarding various properties, where red symbolizes low accuracy and white indicates high accuracy. The delineation between the training and prediction phases is visually represented; a black outline demarcates the training phase, while a blue outline with shading indicates the prediction phase. (b) The influence of computational power on expediting the physical chemistry process is graphically represented. Experimental stages are depicted by blue bars, the model training phase by an orange bar, and the prediction phase by pink bars. Spirals indicate time progression, with each full cycle corresponding to a 4-minute interval, sequentially depicting the stages of experimentation, model training, and outcome prediction.

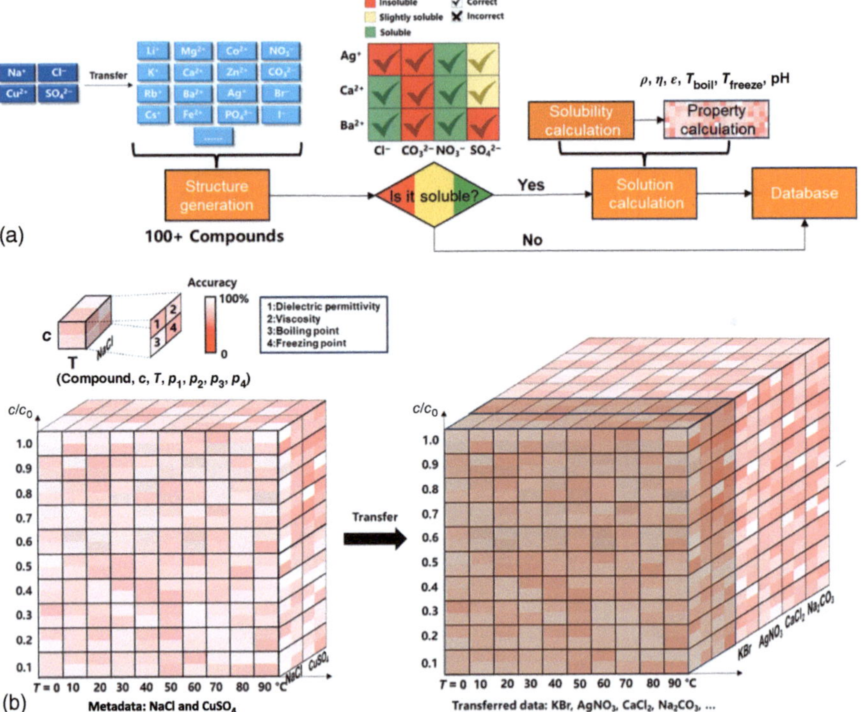

Figure 7.9 Property calculation of general aqueous solution. (a) Scaling scheme to general solutions. Inset Table: Solubility judgment calculation benchmark for compound combinations among Ag, Ca, Ba, chloride, carbonate, nitrate, and sulfate. Then properties (ρ: density, η: viscosity, ε: dielectric constant, T_{boil}: boiling point, T_{freeze}: freezing point) of solution are calculated. b, Accuracy heat map for several ionic compounds (NaCl, CuSO$_4$, KBr, AgNO$_3$, CaCl$_2$, Na$_2$CO$_3$). For each compound, a 2×2 grid of properties is calculated where the color maps the accuracy at a certain temperature and concentration gradient.

model to a broader range of salt solutions. This expansion is depicted in Figure 7.9, illustrating our model's capability to encompass over 100 ionic compounds, totaling more than 10^4 entries. We begin by assessing each compound's solubility in water – categorized as soluble, slightly soluble, or insoluble. For those not deemed insoluble, we further explore the temperature-dependent solubility. Given specific temperatures and concentrations, the model can promptly calculate the solution's physicochemical properties. All resultant data are meticulously evaluated and documented within the metaverse.

The determination of a salt's solubility involves a comparison between its lattice energy (E_{lat}), calculated via the Born–Lande equation – which factors in the Madelung constant, ionic charges, interionic distance, and the Born exponent – and its hydration energy (E_{sol}), estimated using Widom's test particle insertion method. These calculations are heavily reliant on the lattice geometry and elemental species, underscored by spatial and charge parameters. To fully study the correlation between properties, we include all the parameters into a correlation matrix Λ:

$$\Lambda = \begin{bmatrix} E_{\text{lat}} & \lambda_{12} & \lambda_{13} & \cdots & \lambda_{1n} \\ \lambda_{21} & E_{\text{sol}} & \lambda_{23} & \cdots & \lambda_{2n} \\ \lambda_{31} & \lambda_{32} & q & \cdots & \lambda_{3n} \\ \vdots & \vdots & \vdots & \ddots & \vdots \\ \lambda_{n1} & \lambda_{n2} & \lambda_{n3} & \cdots & r_0 \end{bmatrix} \tag{7.4}$$

where λ_{ij} denotes the correlations between each pair of parameters. The initial λ_{ij} parameters are fit from 1-D equations, and we optimize λ_{ij} such that the eigenvalue of E_{lat} and E_{sol} matches the historical values. Then we tested our parameters of calculating solubility for numerous compounds, where our model can achieve nearly 100% accuracy, shown in Figure 7.9a. Our data reveals a significant energy barrier for insoluble salts such as AgCl and $BaSO_4$, while a small energy is overcome for slightly soluble salts such as $CaSO_4$ and Ag_2SO_4. The solvation process of these compounds can be accelerated through external inference such as shaking or heating, creating large energy fluctuations.

Our study focuses on the solubility of various salts in water, highlighting the balance between Coulomb interactions and hydrogen bonds, and how this balance shifts with temperature changes, affecting the hydration of ions. Notably, the solubility of Na_2SO_4 exhibits a non-linear relationship with temperature, influenced by ion cluster sizes and hydrogen bond dynamics, consistent with experimental findings (Bharmoria et al. 2014). Our model successfully predicts these temperature-dependent solubility patterns by accurately modeling hydrogen bond interactions.

We further investigated solution properties such as density, color, and pH, which reflect the underlying ion–water interactions. The model incorporates Pauling's ionic character theory(Pauling 1960) to explain color variations, offering a novel approach to predict solution color based on electronegativity. Hydrolysis, a key factor in pH levels, is analyzed through radial density function analysis to observe H_3O^+ and OH^- formations, crucial for accurately calculating the pH of salts prone to significant hydrolysis.

Additionally, we calculated other physicochemical properties, including viscosity and dielectric constant, across various concentrations and temperatures, presenting our findings in a comprehensive dataset detailing over 100 properties of inorganic salt solutions. This dataset supports the model's applicability to diverse compounds, despite challenges in accurately detecting neutral solution ions and the impact of complex electronic structures on computation accuracy. We aim to extend our model's application to organic systems, leveraging its capability to encode both intra- and intermolecular interactions. This extension will facilitate advancements in crystal and medicine design, incorporating frameworks for crystal structure generation and stability calculations. Optimizing the force field tensor will enable us to encompass both kinetics and thermodynamics, offering a robust tool for industry applications.

7.6 Challenges and Future Directions

Despite the promising future of metaverse lab, a potential hurdle is the complexity involved in integrating a wide range of computational tools and experimental techniques into a cohesive metaverse framework. This necessitates extensive efforts toward standardization and development to ensure compatibility and interoperability across various scientific software and data formats. Moreover, the paramount importance of data privacy and security in the metaverse, where sensitive research data abound, calls for the development of robust encryption methods and secure access protocols to safeguard intellectual property and personal data against unauthorized access and cyber threats. Another challenge is scalability and resource management, as the infrastructure must be scaled efficiently to support more complex simulations and experiments without compromising performance. Ensuring the metaverse remains an intuitive and accessible platform for researchers across different fields is crucial for its widespread adoption, necessitating the creation of user-friendly interfaces.

Looking ahead, the integration of advanced artificial intelligence (AI) and machine learning (ML) models within the metaverse holds the promise of enhancing predictive accuracy and automating data analyses, opening up new frontiers for discovery and innovation. The metaverse's unique capacity for facilitating global collaboration offers researchers the opportunity to work together in real time, transcending physical boundaries. This collaborative ecosystem is expected to shorten research cycles and encourage interdisciplinary innovations. Additionally, the ability to prototype and test hypotheses virtually presents a cost-effective and environmentally friendly alternative to traditional research methods, potentially revolutionizing the R&D process. The metaverse also emerges as a potent educational tool, offering immersive learning experiences that can democratize access to hands-on research and training opportunities. Despite the challenges in realizing the full potential of integrating computational and experimental research in the metaverse, the future looks bright. Continued technological advancements and collaborative efforts within the scientific community are likely to cement the metaverse as a pivotal platform in future research methodologies.

References

Abascal, J.L.F., Sanz, E., García Fernández, R., and Vega, C. (2005). A potential model for the study of ices and amorphous water: TIP4P/Ice. *The Journal of Chemical Physics* 122 (23): 234511.

Abiri, R., Borhani, S., Sellers, E.W. et al. (2019). A comprehensive review of EEG-based brain–computer interface paradigms. *Journal of Neural Engineering* 16 (1): 011001.

Aryanpour, M., van Duin, A.C.T., and Kubicki, J.D. (2010). Development of a reactive force field for iron−oxyhydroxide systems. *The Journal of Physical Chemistry. A* 114 (21): 6298–6307. https://doi.org/10.1021/jp101332k.

Back, S., Aspuru-Guzik, A., Ceriotti, M. et al. (2024). Accelerated chemical science with AI. *Digital Discovery* 3: 23–33.

Batatia, I., Benner, P., Chiang, Y. et al. (2023). *A Foundation Model for Atomistic Materials Chemistry.* arXiv preprint arXiv:2401.00096.

Bauer, J., Spanton, S., Henry, R. et al. (2001). Ritonavir: an extraordinary example of conformational polymorphism. *Pharmaceutical Research* 18 (6): 859–866.

Bharmoria, P., Gehlot, P.S., Gupta, H., and Kumar, A. (2014). Temperature-dependent solubility transition of Na_2SO_4 in water and the effect of NaCl therein: solution structures and salt water dynamics. *The Journal of Physical Chemistry. B* 118 (44): 12734–12742. https://doi.org/10.1021/jp507949h.

Broo, A. and Nilsson Lill, S.O. (2016). Transferable force field for crystal structure predictions, investigation of performance and exploration of different rescoring strategies using DFT-D methods. *Acta Crystallographica. Section B: Structural Science, Crystal Engineering and Materials* 72 (4): 460–476.

Car, R. and Parrinello, M. (1985). Unified approach for molecular dynamics and density-functional theory. *Physical Review Letters* 55 (22): 2471–2474. https://doi.org/10.1103/PhysRevLett.55.2471.

Del Rosso, L., Celli, M., Grazzi, F. et al. (2020). Cubic ice Ic without stacking defects obtained from ice XVII. *Nature Materials* 19 (6): 663–668.

Fedkin, M.V., Shin, Y.K., Dasgupta, N. et al. (2019). Development of the ReaxFF methodology for electrolyte−water systems. *The Journal of Physical Chemistry. A* 123 (10): 2125–2141. https://doi.org/10.1021/acs.jpca.8b10453.

Firaha, D., Liu, Y.M., van de Streek, J. et al. (2023). Predicting crystal form stability under real-world conditions. *Nature* 623 (7986): 324–328. https://doi.org/10.1038/s41586-023-06587-3.

Gao, Y., Lu, Y., and Zhu, X. (2022). Mateverse, the future materials science computation platform based on metaverse. *Journal of Physical Chemistry Letters* 14: 148–157.

Gasser, T.M., Thoeny, A.V., Fortes, A.D., and Loerting, T. (2021). Structural characterization of ice XIX as the second polymorph related to ice VI. *Nature Communications* 12 (1): 1–10.

Gerasimov, I.S., Losev, T.V., Epifanov, E.Y. et al. (2022). Comment on "Pushing the frontiers of density functionals by solving the fractional electron problem". *Science* 377: 6606.

Glass, C.W., Oganov, A.R., and Hansen, N. (2006). USPEX—Evolutionary crystal structure prediction. *Computer Physics Communications* 175 (11–12): 713–720.

Harris, K.R. and Woolf, L.A. (1980). Pressure and temperature dependence of the self diffusion coefficient of water and oxygen-18 water. *Journal of the Chemical Society, Faraday Transactions 1* 76: 377–385.

Hoffmann, R. and Malrieu, J.P. (2020). Simulation vs. understanding: a tension, in quantum chemistry and beyond. Part A. Stage setting. *Angewandte Chemie, International Edition* 59 (31): 12590–12610.

Hollingsworth, S.A. and Dror, R.O. (2018). Molecular dynamics simulation for all. *Neuron* 99 (6): 1129–1143.

Ignatov, I. and Mosin, O. (2014). Nature of hydrogen bonds in liquids and crystals. Ice crystal modifications and their physical characteristics. *Journal of Medicine, Physiology and Biophysics* 4: 58–80.

Jorgensen, W.L., Chandrasekhar, J., Madura, J.D. et al. (1983). Comparison of simple potential functions for simulating liquid water. *The Journal of Chemical Physics* 79 (2): 926–935.

Jorgensen, W.L. and Madura, J.D. (1985). Temperature and size dependence for Monte Carlo simulations of TIP4P water. *Molecular Physics* 56 (6): 1381–1392.

Kadaoluwa Pathirannahalage, S.P., Meftahi, N., Elbourne, A. et al. (2021). Systematic comparison of the structural and dynamic properties of commonly used water models for molecular dynamics simulations. *Journal of Chemical Information and Modeling* 61 (9): 4521–4536.

Karplus, M. and McCammon, J.A. (2002). Molecular dynamics simulations of biomolecules. *Nature Structural Biology* 9 (9): 646–652.

Khalak, Y., Baumeier, B., and Karttunen, M. (2018). Improved general-purpose five-point model for water: TIP5P/2018. *The Journal of Chemical Physics* 149 (22): 224507.

Kingma, D.P. and Welling, M. (2013). *Auto-Encoding Variational Bayes*. arXiv preprint arXiv:1312.6114.

Kirkpatrick, J., McMorrow, B., Turban, D.H.P. et al. (2021). Pushing the frontiers of density functionals by solving the fractional electron problem. *Science* 374 (6573): 1385–1389.

Kirkpatrick, J., McMorrow, B., Turban, D.H.P. et al. (2022). Response to Comment on "Pushing the frontiers of density functionals by solving the fractional electron problem". *Science* 377: 6606.

Kofke, D.A. (1993). Direct evaluation of phase coexistence by molecular simulation via integration along the saturation line. *The Journal of Chemical Physics* 98 (5): 4149–4162.

Li, J., Li, J., Liu, R. et al. (2020). Autonomous discovery of optically active chiral inorganic perovskite nanocrystals through an intelligent cloud lab. *Nature Communications* 11 (1): 2046.

Li, T. (2007). Understanding the polymorphism of aspirin with electronic calculations. *Journal of Pharmaceutical Sciences* 96 (4): 755–760.

Lyakhov, A.O., Oganov, A.R., Stokes, H.T., and Zhu, Q. (2013). New developments in evolutionary structure prediction algorithm USPEX. *Computer Physics Communications* 184 (4): 1172–1182.

Makkar, P. and Ghosh, N.N. (2021). A review on the use of DFT for the prediction of the properties of nanomaterials. *RSC Advances* 11 (45): 27897–27924.

Malmberg, C.G. and Maryott, A.A. (1956). Dielectric constant of water from 0°C to 1000°C. *Journal of Research of the National Bureau of Standards* 56 (1): 1–8.

Mayo, S.L., Olafson, B.D., and Goddard, W.A. (1990). DREIDING: a generic force field for molecular simulations. *The Journal of Physical Chemistry* 94 (26): 8897–8909.

Merchant, A., Batzner, S., Schoenholz, S.S. et al. (2023). Scaling deep learning for materials discovery. *Nature* 624 (7990): 80–85. https://doi.org/10.1038/s41586-023-06735-9.

Millot, M., Coppari, F., Rygg, J.R. et al. (2019). Nanosecond X-ray diffraction of shock-compressed superionic water ice. *Nature* 569 (7755): 251–255.

Monti, S., Li, C., and Carravetta, V. (2013). Reactive dynamics simulation of monolayer and multilayer adsorption of glycine on Cu(110). *Journal of Physical Chemistry C* 117 (10): 5221–5228. https://doi.org/10.1021/jp312828d.

Musk, E. (2019). An integrated brain-machine interface platform with thousands of channels. *Journal of Medical Internet Research* 21 (10): e16194.

Park, H., Onwuli, A., Butler, K.T., and Walsh, A. (2024). *Mapping Inorganic Crystal Chemical Space*. ChemRxiv.

Pauling, L. (1960). *The Nature of Chemical Bond—An Introduction to Modern Structural Chemistry*, 3e. Ithaca, New York: Cornell University Press.

Polian, A. and Grimsditch, M. (1984). New high-pressure phase of H_2O: ice X. *Physical Review Letters* 52 (15): 1312.

Rahaman, O., van Duin, A.C.T., Bryantsev, V.S. et al. (2010). Development of a ReaxFF reactive force field for aqueous chloride and copper chloride. *The Journal of Physical Chemistry. A* 114 (10): 3556–3568. https://doi.org/10.1021/jp9090415.

Rezeika, A., Benda, M., Stawicki, P. et al. (2018). Brain–computer interface spellers: a review. *Brain Sciences* 8 (4): 57.

Ronneberger, O., Fischer, P., & Brox, T. (2015). U-net: convolutional networks for biomedical image segmentation. Paper presented at the International Conference on Medical Image Computing and Computer Assisted Intervention.

Salzmann, C.G., Loveday, J.S., Rosu-Finsen, A., and Bull, C.L. (2021). Structure and nature of ice XIX. *Nature Communications* 12 (1): 1–7.

Shaw, D. E., Adams, P. J., Azaria, A., et al. (2021). Anton 3: twenty microseconds of molecular dynamics simulation before lunch. Paper presented at the The International Conference for High Performance Computing, Networking, Storage, and Analysis.

Shtukenberg, A.G., Hu, C.T., Zhu, Q. et al. (2017). The third ambient aspirin polymorph. *Crystal Growth & Design* 17 (6): 3562–3566.

Wu, X., Xiao, L., Sun, Y. et al. (2022). A survey of human-in-the-loop for machine learning. *Future Generation Computer Systems* 135: 364–381.

Yann, L., Bengio, Y., and Hinton, G. (2015). Deep learning. *Nature* 521 (7553): 436–444.

Yann, L., Touresky, D., Hinton, G., & Sejnowski, T. (1988). A theoretical framework for back-propagation. Paper presented at the Proc. 1988 Connectionist Models Summer School.

Zeron, I., Abascal, J., and Vega, C. (2019). A force field of Li^+, Na^+, K^+, Mg^{2+}, Ca^{2+}, Cl^-, and SO_4^{2-} in aqueous solution based on the TIP4P/2005 water model and scaled charges for the ions. *The Journal of Chemical Physics* 151 (13): 134504.

Zhang, L., Han, J., Wang, H. et al. (2018). Deep potential molecular dynamics: a scalable model with the accuracy of quantum mechanics. *Physical Review Letters* 120 (14): 143001. https://doi.org/10.1103/PhysRevLett.120.143001.

Zoete, V., Cuendet, M.A., Grosdidier, A., and Michielin, O. (2011). SwissParam: a fast force field generation tool for small organic molecules. *Journal of Computational Chemistry* 32 (11): 2359–2368.

Index

a

AbsoluteProof software   139
advanced automated synthesis platform
        42, 45
AI-assistant chemical and bio-material
        design   58
AI-driven automated microscopy system
        42
AI-powered digital chemical synthesis
        platform   40, 42
AI prediction   40, 166, 171–172
AI Supervisor and ScholarNet   130–134
AlexNet   73, 85
AlphaFlow   48–52
AlphaFold   165
AnByCz   118
Apple Vision Pro   164
aqueous system research in metaverse
        177–181
Arrhenius equation   4
automated chemical reactor system   25
automated chemical synthesis systems
        26, 27
automated intelligent machine (AIM)   47
autonomous high throughput
        experiments   166–172

b

band theory   6
Bayesian neural networks (BNNs)   92
BiaeP   150–155, 157
Big Kahuna System   36, 37, 42
BindingDB   71

Bitcoin   140–142
"black-box" optimization algorithms   79
BlockAudit   144
blockchain in the autonomous laboratory
        150–155
blockchain technology   139–145, 150,
        157–160, 163
Born–Lande equation   180
Byzantine fault tolerance (BFT)   140, 144
Byzantine Generals Problem   140, 141

c

canonical SMILES   60–63
ChatGPT   19, 114, 131
CHEMBL   70
ChemOS   96
Chempiler program   91
Chemputer system   91, 92
components identification   101–104
computer-aided synthesis planning
        (CASP)   40, 94, 95
convergence innovation   129
convolutional neural networks (CNNs)
        72–74, 89
coolant and moderator materials   10
cooling bath system   23, 24
CrossRef   121
CubeRoot   155–157

d

data balancing   68
dataset preprocessing   68–69
data storage safety   158–159

AI and Robotic Technology in Materials and Chemistry Research, First Edition. Xi Zhu.
© 2025 WILEY-VCH GmbH. Published 2025 by WILEY-VCH GmbH.

decentralized applications (dApps) 142
decentralized autonomous organizations
(DAOs) 142
decentralized finance (DeFi) 142
DeepCID 99
DeepMind 165
deep neural networks (DNNs) 72, 77, 85,
99
DeepRiPP workflow 81
digital chemical synthesis platform 39,
40, 42
digital currencies 142, 163
digitalization 29, 165–167
divergence innovation 129

e

Edisonian design 78
Emerald Cloud Lab (ECL) 35, 36, 42
Ethereum 142, 143, 150–154
Ethereum Blockchain 142, 143, 150, 152,
153
Ethereum Blockchain & Cloud Storage
Platform 150
Ethereum Virtual Machine (EVM) 142

f

first-principle-based inverse design
78–79
flow-based automated reaction device
91
Foldit 165
Fortnite 164
fuel cladding materials 10

g

Gaussian approximation potential (GAP)
76, 77
Gaussian process (GP) 95
Generalized Gradient Approximation
(GGA) 67
Generative Pre-trained Transformer
(GPT) 117
Gibbs–Duhem integration methods 175
GlobusLabs 116

gradient-domain machine learning
(GDML) approach 76
graph attention networks (GATs)
74–75
graph convolutional neural networks
(GCN) 73–75
graph neural networks (GNNs) 65, 71,
74
Graph-pKa 66

h

hard drive-based PoW systems 141
high-performance computing devices
164
Hohenberg-Kohn theorems 67
HTC Vive 164
human-in-loop (HIL) optimization
165–166, 169–171, 173
human-refined multistep synthesis 94
hydrolysis 177, 181
hyper-converged Autonomous Organic
Reaction Infrastructure (HAORI)
96, 97

i

IBM RXN for Chemistry (IBM) 39, 92
ice phase diagram 175
image captioning (IC) 114
ImageNet 120, 130
in-context learning (ICL) 116, 117, 119
Industry 5.0 163–165
Inferred Contextual Learning (ICL) 116,
119
intelligent machines 19
International Chemical Identifier (InChI)
58, 59, 62–63
International Chemical Identifier Key
(InChIKey) 63
isomeric SMILES 60

k

Kapok series of advanced automated
laboratory equipment 49
Kohn–Sham equation 66–68, 75–78

l

laboratory information management
systems (LIMS)   29
LangChain's ConversationChain method
121
Language-Interfaced Fine-Tuning (LIFT)
approach   114, 117
framework   117
large language models (LLMs)   104
Large Scale Visual Recognition Challenge
(LSVRC)   85
linear notations for molecules   58–63
Local Density Approximation
(LDA)   67

m

machine learning (ML)   21, 29, 36, 58,
68, 69, 71, 72, 75, 76, 78, 80, 81, 89,
94, 95, 99, 108, 114, 182
MAPI-LLM framework   118
materials acceleration operation system
(MAOS)   166
message-passing neural networks
(MPNN)   75–76
metal-organic frameworks (MOFs)   34,
35, 146
Microsoft HoloLens   164
MIDA boronate-based purification
method   91
molecular graph representation   63
molecular representation and encoding
58–66
molecular structure predictor   102, 103

n

Natural Language Processing (NLP)   80,
99, 104, 113, 114, 119
Neuralink   169
"no free lunch" theorems   79
non-fungible tokens (NFTs)   163

o

object-relational mapping (ORM)   144
Oculus Rift   164
OpenAlex   121

optimization algorithms   78, 79, 95, 176
organic semiconductor lasers
(OSLs)   95

p

Partial Least Squares (PLS)   98
Pauling's ionic character theory   181
physics-endorsed diffusion-like model
171
physics-inspired neural networks   77
PlayStation VR   164
practical Byzantine fault tolerance (pBFT)
140
principal component analysis (PCA)   98
process analytical technology (PAT)   94
Protein Data Bank (PDB) database   70,
71
PubChem   70, 71
PubMed   121
PubPeer   146–148

q

quantum deep field (QDF)   77

r

reinforcement learning (RL)   48, 51,
89–90
renewable energy   7–10, 110
ResNet   73
Roblox   164
RoboChem   97
RoboRXN from IBM   92
Robot Scientist Adam   31, 42

s

Savitzky–Golay filter   101
SchNet   75–77
ScholarBert model   116
ScholarNet   120, 130–134
scite   114
self-referencing embedded strings
(SELFIES)   116
Simplified Molecular Input Line Entry
System (SMILES)   58–63, 70, 116

SMILES Arbitrary Target Specification
(SMARTS) 59, 62
Soft Independent Modeling of Class
Analogy (SIMCA) 98
Spectrum Spiking Neural Network
(SpecSNN) architecture
96, 97
Spider Matrix 120, 124–130, 136, 168
Symyx Tools 31, 32, 42

t

tensor networks (TNs) 99–100, 120
TIP4P-Meta 174–176

u

Unchained Labs' Big Kahuna System 36
U-net architecture 171

Unity 13, 164
Unreal Engine 164

v

variational autoencoder (VAE) 171
Vision and Language BERT (ViLBERT)
114
Vision Transformer (ViT) 114
VisualBERT 114

w

water force field TIP4P-Meta 174
wearable technology 164
Word2Vec 113

z

Ziegler–Natta catalyst 15